Water Harvesting in Sub-Saharan Africa

T0199853

Agriculture in Sub-Saharan Africa is constrained by highly variable rainfall, frequent drought and low water productivity. There is an urgent need, heightened by climate change, for appropriate technologies to address this problem through managing and increasing the quantity of water on farmers' fields – by water harvesting. This book defines water harvesting as a set of approaches which occupy an intermediate position along the water-management spectrum extending from *in situ* moisture conservation to irrigated agriculture. They generally comprise small-scale systems that induce, collect, store and make use of local surface runoff or floodwaters for agriculture.

The authors review development experience and set out the state of the art of water harvesting for crop production and other benefits in Sub-Saharan Africa. This includes an assessment of water harvesting schemes that were initiated two or three decades ago when interest was stimulated by the droughts of the 1970s and 1980s. These provide lessons to promote sustainable development of dryland agriculture in the face of changing environmental conditions. Case studies from eight countries across western, central, eastern and southern Africa provide the evidence base. Each follows a similar format and is based on assessments conducted in collaboration with in-country partners, with a focus on attempts to promote adoption of water harvesting, both horizontally (spread) and vertically (institutionalization). Introductory cross-cutting chapters as well as an analytical conclusion are also included.

William Critchley is a Senior Sustainable Land Management Specialist at the Centre for International Cooperation, VU University Amsterdam, the Netherlands.

John Gowing is a Reader in Agricultural Water Management, School of Agriculture, Food and Rural Development, Newcastle University, Newcastle upon Tyne, UK.

About CTA

The Technical Centre for Agricultural and Rural Cooperation (CTA) is a joint international institution of the African, Caribbean and Pacific (ACP) Group of States and the European Union (EU). Its mission is to advance food and nutritional security, increase prosperity and encourage sound natural resource management in ACP countries. It provides access to information and knowledge, facilitates policy dialogue and strengthens the capacity of agricultural and rural development institutions and communities.

CTA operates under the framework of the Cotonou Agreement and is funded by the EU. For more information on CTA, visit www.cta.int.

Water Harvesting in Sub-Saharan Africa

Edited by

William Critchley and John Gowing

Associate Editor: Eefke Mollee

LONDON AND NEW YORK

from Routledge

This first edition published 2012
by Routledge
2 Park Square, Milton Park, Abingdon, Oxon OX14 4RN

Simultaneously published in the USA and Canada
by Routledge
711 Third Avenue, New York, NY 10017

Routledge is an imprint of the Taylor & Francis Group, an informa business

British Library Cataloguing in Publication Data
A catalogue record for this book is available from the British Library

Library of Congress Cataloging-in-Publication Data
Water harvesting in Sub-Saharan Africa / edited by William Critchley and
John Gowing.
    p. cm.
Includes bibliographical references and index.
1. Water harvesting—Africa, Sub-Saharan. 2. Water-supply—Africa,
Sub-Saharan. I. Critchley, Will. II. Gowing, John W.
TD418.W36 2012
627'.50967—dc23   2012024623

ISBN13: 978-0-415-53773-5 (hbk)
ISBN13: 978-0-415-53786-5 (pbk)
ISBN13: 978-0-203-10998-4 (ebk)

Typeset in Times
by Cenveo Publisher Services

Contents

Contributors

Abebe Awass A. Arba Minch University, Institute of Technology (AMIT), P.O. Box 21, Arba Minch, Ethiopia.

Alemu E. Arba Minch University, Institute of Technology (AMIT), P.O. Box 21, Arba Minch, Ethiopia.

Balima M. National Institute for Environment and Agricultural Research (INERA), 04 BP 7192 Ouagadougou, Burkina Faso.

Barron J. University of York, Stockholm Environment Institute (SEI), Heslington, York, YO10 5DD, UK and Stockholm Resilience Centre, Stockholm University.

Bouma J.A. VU University Amsterdam, Institute for Environmental Studies (IVM), De Boelelaan 1087, 1081 HV Amsterdam, the Netherlands.

Bwalya M. NEPAD Planning and Coordinating Agency – CAADP , P.O. Box 1234, Halfway House, Midrand, Johannesburg, South Africa.

Cherogony K.R. World Agroforestry Centre (ICRAF), United Nations Avenue, Gigiri, P.O. Box 30677, Nairobi, 00100, Kenya.

Critchley W.R.S. VU University Amsterdam, Centre for International Cooperation (CIS), De Boelelaan 1105, 1081 HV Amsterdam, the Netherlands.

Deen A.M. Land Use and Desertification Control (LUDC) department, Kassala State, Sudan.

Di Prima S.L. VU University Amsterdam, Centre for International Cooperation (CIS), De Boelelaan 1105, 1081 HV Amsterdam, the Netherlands.

Eissa A.O. Soil Conservation, Land Use and Water Programming Administration (SCLUWPA), Port Sudan, Sudan.

Gaiballa A.K. Sudan University of Science and Technology, College of Forestry and Range Science (SUST-CFRS), P.O. Box 407, Khartoum, Sudan.

Gowing J.W. Newcastle University, School of Agriculture, Food & Rural Development, Newcastle upon Tyne, NE1 7RU, UK.

Gumbo D. University of Zimbabwe, Department of Soil Science and Agricultural Engineering, Faculty of Agriculture, P.O. Box MP 167, Mount Pleasant, Harare, Zimbabwe.

Hassane A. Independent agronomist, B.P. 13 191, Niamey, Niger.

Kahimba F.C. Sokoine University of Agriculture, Faculty of Agriculture, Department of Agricultural Engineering and Land Planning, P.O. Box 3003, Morogoro, Tanzania.

Lasage R. VU University Amsterdam, Institute for Environmental Studies (IVM), De Boelelaan 1087, 1081 HV Amsterdam, the Netherlands.

Mahoo H.F. Sokoine University of Agriculture, Faculty of Agriculture, Department of Agricultural Engineering and Land Planning, P.O. Box 3003, Morogoro, Tanzania.

Malesu M. Southern and Eastern Africa Rainwater Network (SearNet), World Agroforestry Centre (ICRAF), United Nations Avenue, Gigiri, P.O. Box 30677, Nairobi, 00100, Kenya.

Mbilinyi B.P. Sokoine University of Agriculture, Faculty of Agriculture, Department of Agricultural Engineering and Land Planning, P.O. Box 3003, Morogoro, Tanzania.

Mollee E.M. VU University Amsterdam, Centre for International Cooperation (CIS), De Boelelaan 1105, 1081 HV Amsterdam, the Netherlands.

Mutabazi K.D. Sokoine University of Agriculture, Faculty of Agriculture, Department of Agricultural Economics and Agribusiness, P.O. Box 3007, Morogoro, Tanzania.

Nyagumbo I. University of Zimbabwe, Department of Soil Science and Agricultural Engineering, Faculty of Agriculture, P.O. Box MP 167, Mount Pleasant, Harare, Zimbabwe. Also CIMMYT-International Maize and Wheat Improvement Centre, Southern Africa Regional Office, P.O. Box MP 163, Mt. Pleasant, Harare, Zimbabwe.

Oduor A.R. Southern and Eastern Africa Rainwater Network (SearNet), World Agroforestry Centre (ICRAF), United Nations Avenue, Gigiri, P.O. Box 30677, Nairobi, 00100, Kenya.

Ouattara K. National Institute for Environment and Agricultural Research (INERA), 04 BP 7192 Ouagadougou, Burkina Faso.

Ouédraogo I. National Institute for Environment and Agricultural Research (INERA), 04 BP 7192 Ouagadougou, Burkina Faso.

Oughton E. Newcastle University, School of Agriculture, Food & Rural Development, Newcastle upon Tyne, NE1 7RU, UK.

Reij C. VU University Amsterdam, Centre for International Cooperation (CIS), De Boelelaan 1105, 1081 HV Amsterdam, the Netherlands.

Reuben P. Sokoine University of Agriculture, Faculty of Agriculture, Department of Agricultural Engineering and Land Planning, P.O. Box 3003, Morogoro, Tanzania.

Rwehumbiza F.B. Sokoine University of Agriculture, Faculty of Agriculture, Department of Soil Science, P.O. Box 3008, Morogoro, Tanzania.

Sally H. Independent irrigation and water management specialist, Sri Lanka.

Savadogo M. National Institute for Environment and Agricultural Research (INERA), 04 BP 7192 Ouagadougou, Burkina Faso.

Sawadogo-Kabore S. National Institute for Environment and Agricultural Research (INERA), 04 BP 7192 Ouagadougou, Burkina Faso.

Scheierling S.M. World Bank, 1818 H Street, NW Washington, DC 20433, United States.

Snelder D.J.R.M. VU University Amsterdam, Centre for International Cooperation (CIS), De Boelelaan 1105, 1081 HV Amsterdam, The Netherlands and Leiden University, Institute of Environmental Sciences (CML).

Traoré S. National Institute for Environment and Agricultural Research (INERA), 04 BP 7192 Ouagadougou, Burkina Faso.

Tumbo S.D. Sokoine University of Agriculture, Faculty of Agriculture, Department of Agricultural Engineering and Land Planning, P.O. Box 3003, Morogoro, Tanzania.

Woldearegay Woldemariam K. Mekelle University, Tigray Regional State, 231, Mekelle, Ethiopia.

Wuta M. CIMMYT-International Maize and Wheat Improvement Centre, Southern Africa Regional Office, P.O. Box MP 163, Mt Pleasant, Harare, Zimbabwe.

Acronyms and abbreviations

ADRK	Association for the Development of the Kaya Region
AFVP	Association Française des Volontaires du Progrès
AGRITEX	Department of Agricultural, Technical and Extension Services
AQUASTAT	FAO's global information system on water and agriculture
ASAL	Arid and Semi-Arid Lands
BFFP	Baringo Fuel and Fodder Project
BPSAAP	Baringo Pilot Semi-Arid Area Project
CFA	Currency used in several West African countries
CIEPAC	Centre International pour l'Education Permanente et l'Aménagement Concerté
CILSS	Comité permanent Inter-Etats de Lutte contre la Sécheresse au Sahel
CPPSLM	Country Partnership Programme for Sustainable Land Management
CRPA	Centre Régional de Promotion Agro-Pastorale
DLC	Dead-level contour
DPDAD	Déclaration de Politique de Développement Agricole Durable
DR&SS	Department of Research and Specialist Services
ERDP	Eastern Recovery Development Programme
ERHA	Ethiopian Rainwater Harvesting Association
EthiOCAT	Ethiopian Overview of Conservation Approaches and Technologies
FAEP	Fuelwood Afforestation Extension Programme
FAO	Food and Agriculture Organization of the United Nations
FDR	Fonds de Développement Rural
FDRE	Federal Democratic Republic of Ethiopia
FEER	Fonds de l'Eau et de l'Equipement Rural
FFA	Food-for-assets
FFS	Farmer Field Schools
FFW	Food-for-work
FNGN	Fédération Nationale des Groupements Naam
FTLRP	Fast Track Land Resettlement Program
GDP	Gross domestic product
GERES	Groupement Européen de Restauration des Sols

GIZ	Deutsche Gesellschaft für Internationale Zusammenarbeit (GmbH)
GTZ	Deutsche Gesellschaft für Technische Zusammenarbeit
ICARDA	International Centre for Agricultural Research in Dry Areas
ICRAF	International Centre for Research in Agroforestry (now World Agroforestry Centre)
IFAD	International Fund for Agricultural Development
IFERD	Institute for Research and Training, Education and Development
IK	Indigenous knowledge
ILWRM	Integrated Land and Water Resources Management
INERA	Institut de l'Environnement et de Recherches Agricoles
IPCC	Intergovernmental Panel on Climate Change
ISCO	International Soil Conservation Organisation
ITCZ	Intertropical Convergence Zone
IWRM	Integrated Water Resource Management
KRA	Kenya Rainwater Association
LGU	Local government unit
LSCF	Large-scale commercial farm
LUDC	Land Use and Desertification Control Branch (of the State Ministry of Agriculture)
MAHRH	Ministère de l'Agriculture, de l'Hydrauliquc ct des Ressources Halieutiques (Burkina Faso)
MDG	Millennium Development Goal
MERET	Managing Environmental Resources to Enable Transition
MoA	Ministry of Agriculture
NGO	Non-governmental organisation
NEDECO	Netherlands Engineering Consultants
NEPA	National Environmental Policy of Ethiopia
NIDP	National Irrigation Development Plan
NIMP	National Irrigation Master Plan
ODA	Official development assistance
ONG	Organisation non-gouvernementale
PAE	Projet Agro Ecologie
PAF	Projet Agro-Forestier
PAGIFS	Le Plan de Gestion Intégrée de la Fertilité des Sols
PAPANAM	Le Projet Action de Production et d'Accompagnement dans la province du Namentenga
PASDEP	Plan for Accelerated and Sustained Development to End Poverty
PATECORE	Projet d'Aménagement des Terroirs et de Conservation de Ressources
PEDI	Programme d'Exécution du Développement Intégré
PIDP	Participatory Irrigation Development Programme
PNGT	Programme National de Gestion des Terroirs
PRRO	Protracted Relief and Rehabilitation Operations

PSB	Burkina Faso Sahel Programme
PS-CES/AGF	Programme Spécial de Conservation des Eaux et des Sols et d'Agroforesterie
PSN1/2	Programme Spécial National FIDA NIGER Phase 1/Phase 2
PVNY	Projet Vivrier Nord Yatenga
RELMA	Regional Land Management Unit (East Africa)
RMARP	Réseau Méthodes Accélérées de Recherche Participative
RRH	Road runoff harvesting
RSCU	Regional Soil Conservation Unit
RWH	Rainwater harvesting
SCLUWPA	Soil Conservation, Land Use and Water Programming Administration
SearNet	Southern and Eastern Africa Rainwater Network
Sida	Swedish International Development Authority
SNGFS	Stratégie Nationale de Gestion Intégrée de la Fertilité des Sols SSA Sub-Saharan Africa
SNNPR	Southern Nations, Nationalities, and People's Region
SSI	Smallholder System Innovations
SSWHS	Sub-Saharan Water Harvesting Study
SUA	Sokoine University of Agriculture
SWC/WH	Soil and Water Conservation/Water Harvesting
SWMRG	Soil-Water Management Research Group
TerrAfrica	International partnership addressing land degradation in SSA
TRP	Turkana Rehabilitation Programme
USCAPP	Uganda Soil Conservation and Agroforesty Pilot Project
WARK	Water Spreading Research Kassala
WF	Water Footprint
WFP	World Food Programme
WH	Water harvesting
WH+	Water harvesting plus
WHaTeR	Water Harvesting Technologies Revisited
WOCAT	World Overview of Conservation Approaches and Technologies
WPLL	Western Pare Lowlands
WUA	Water Users Association

Acknowledgements

This book was originally suggested by Tim Hardwick some years ago: he felt there was a need for an update on water harvesting in Africa. His persistence paid off and finally we were unable to resist any longer. Thanks to Tim for his encouragement and we hope that he is pleased with the result. Ashley Irons has acted as a (necessarily) strict and efficient editor, while remaining thoroughly supportive: hers has not been an easy position – wedged between the editors and the production team. Nevertheless she should be deservedly satisfied with how it has all worked out and we are most grateful to her. Appreciation too, to Apurupa Mallik and team for their prompt, polite and eagle-eyed work of uncovering many errors and correcting proofs. On the scientific level this book would not have been possible without the project WHaTeR ('Water Harvesting Technologies Revisited: Potentials for Innovations, Improvements and Upscaling in Sub-Saharan Africa') which is funded through the European Commission's FP7 Instrument. The project began in 2011 with a systematic programme of revisits to countries where water harvesting had been documented 20 to 25 years ago. Much of the information in the country-based chapters, particularly, is derived from that project. We are especially grateful to Mr Balabanis Panagiotis, WHaTeR's Programme Officer. And naturally all those who are involved in this project, many of whom have provided material for the book, are thanked for their (unpaid) contributions. We would also like to thank Gemma Betsema for stepping in, at the last minute, in a voluntary capacity to help most efficiently with editing the chapters. And finally our true appreciation of the work put in by Eefke Mollee whose professional commitment and personal dedication has gone well beyond that expected of an associate editor.

Foreword

This important and timely book addresses one of the most critical issues related to water and food security – enhancing the role of rainfed agriculture. The recently completed Comprehensive Assessment of Water Management in Agriculture, put together by over 1000 scientists and practitioners, concluded that we must invest in better use of rainfall where it limits production. Improved rainfed agriculture is vital to enhance food security and food production in many parts of the world, and to support sustainable development by reducing environmental degradation, and reducing stress on river systems. The book addresses the vital question of *how* to make this happen, and thus provides critical information to practitioners and researchers.

The book takes a retrospective–prospective approach to the topic of rainfed agriculture, and positions water harvesting as a key technique for the drylands. The book demonstrates some of the key opportunities, yet recognizes many of the challenges faced by farmers, practitioners, and policy makers. Some simple technologies such as *zaï* and stone lines in the Sahel have taken off. Others, where heavy machinery was used in projects two decades ago have simply fallen into disrepair. There are yet more cases where introduced systems have evidently worked well but been unattractive to farmers. On balance there is positive news, especially if we learn lessons from the past.

The editors, and many authors, have long experience regarding the state of water harvesting in dryland Africa – and their messages are worth heeding. They point out that it is not just the efficiency of technology that is the issue. Where there has been notable success, it has been founded on simple affordable systems, often based in tradition and supported over the long haul by agencies – commonly by governments themselves. Where there was failure it was to do with hasty and short-lived project interventions in the post-drought 1980s, using food-for-work, heavy machinery and inappropriate systems without adequate forethought. Thus, the way water harvesting interventions have been implemented is as important as the design of the technologies themselves. It is uncanny that a number of these points had been predicted a quarter a century ago, yet many of the same mistakes are perpetuated today.

As the need for more rational use of agricultural water becomes ever more imperative in the semi-arid zones of Africa – with growing populations and climate change becoming more than an uncomfortable reality – water harvesting will increasingly be turned to as a lifeline for millions. It is a climate change adaptation measure of huge potential, yet based on a track-record of thousands of years. While the editors make it clear that this is no design manual, there is plenty of detail to inform water harvesting specialists. That information is, however, holistic and encapsulated by the notion of "water harvesting plus" where society and governance are as much part of the overall solution as hydrology and plant water requirements. They plead for water harvesting to be championed once again – as it was in the 1980s – but warn against simplistic claims of a new panacea. They argue that there is little time to waste; and while more research is certainly valuable, enough is known to go forward with confidence. That surely is an important message.

David Molden

Director General
International Centre for Integrated Mountain Development

Editor: Water for Food, Water for Life: A Comprehensive Assessment of Water Management in Agriculture
IWMI, Earthscan, 2007

Chapter 1

Introduction

William Critchley and John Gowing

This book marks the renaissance of interest in the potential of water harvesting for plant production in Sub-Saharan Africa. The first wave of attention was triggered by the widespread droughts of the 1970s and 1980s, and concerns about 'desertification'. How could crop performance be ensured where rainfall was scarce, and apparently on a downward trend, and droughts increasingly common? Could the key possibly be rainfall runoff that was simultaneously being lost? Development practitioners looked with curiosity and optimism towards experience in the Negev desert in Israel (e.g. Evanari et al., 1971), and also to the USA (e.g. Dutt et al., 1981). They also began to recognize that there were traditions of water harvesting in Africa that could, perhaps, be built upon (Pacey and Cullis, 1986). Explorations through the literature even showed that there had been sporadic attempts to harness runoff waters in previous decades (e.g. in Kenya; Fallon, 1963). An array of initiatives were undertaken throughout West and East Africa in the 1980s – often driven by non-governmental organizations (NGOs) and supported through drought-relief programmes. However these experiments (as many were) and experiences were occurring independently, and there was little sharing of knowledge or analysis of impact. The linguistic divide between francophone and anglophone Africa was an added barrier to exchange of understanding. For this reason, in 1987, the World Bank set out to establish the 'Sub-Saharan Water Harvesting Study'. A further objective of the study was to uncover indigenous practices that were hitherto undocumented, or even unrecognized. A baseline literature review was first produced (Reij et al., 1988) and then, after the field study, the results were reported and analysed (Critchley et al., 1992).

The Sub-Saharan Water Harvesting Study (SSWHS) presented 13 case studies; two of them were entirely traditional, two were project-implemented designs based on traditions, and the remainder designed and implemented by projects (Critchley et al., 1992). The analysis of these cases came up with a series of recommendations, many of which will be familiar to those reading the report 20 years later. They included the need to look into reasons for adoption and non-adoption, the potential gains from sharing information between projects, the need to beware of dependence on heavy machinery, and the suggestion to coordinate incentive mechanisms within countries (Critchley et al., 1992). In many ways this current book takes that SSWHS report as its reference point, and sets out to record what we have learnt since, and where we go from here. That indeed is the basis for the ongoing project, WHaTeR that has led to the collection of much of the data analysed in the 'country chapters' that follow.[1]

It would be incorrect to say that water harvesting has been forgotten in Sub-Saharan Africa since the first phase of activities a quarter of a century ago. But it is surely true that interest has been rekindled in the last few years. That has resulted from the convergence of three current concerns: the potential impacts of climate change in dry areas, increasingly limited water for agriculture, and the imperative of feeding a growing population. Thus water harvesting is firmly back on the mainstream agenda for the drylands of Africa.

It is well to introduce a word of caution from the outset. Water harvesting is no panacea; it is not a silver-bullet solution to Sub-Saharan Africa's problems. Too often in the last few decades, much-heralded 'answers' have led to disappointment at best: 'wonder crops' such as jojoba and tepary bean, exotic tree species such as *Prosopis juliflora*, the one-size-fits-all 'training and visit' system of extension, vetiver grass as the ubiquitous barrier to erosion, and most recently jatropha as a versatile biofuel cash-crop. But there are no such simplistic answers, and hype doesn't help.

Water harvesting delivers benefits by magnifying runoff. What it does *best* is to increase water supplies to crops when they are at a sensitive growth stage – or to tide them over drought spells – and in these situations their performance can be dramatically improved (see Photo 1.1 for an example of a 'contour ridge' technology from Kenya). Borrowing economic terminology, a marginal increase in available water gives the maximum incremental yield. But it is surely self-evident that if the rains fail and drought is prolonged, then water harvesting fails with it. Equally, if rainfall is especially intense, then water harvesting will potentially increase the potency of damaging runoff. Water harvesting has a compelling role to play, but as just one part of an overall strategy of sustainable land management.

This book is not a design manual on water harvesting systems. There are plenty of these available for all different types of water harvesting – the latest is an excellent volume by Oweis and colleagues (2012) which goes into great detail regarding hydrology and engineering design for a wide range of systems. With respect to spate irrigation, Steenbergen et al. (2010) have set the standard with their FAO guidelines. The construction of sand dams is covered by a 'practical guide' written by the Rainwater Harvesting Implementation Network (undated; but this century). One of the current editors co-authored FAO's early design manual for water harvesting (Critchley and Siegert, 1991). Before that, Nissen-Pieterson (1982) had written the standard design manual on rooftop harvesting for Africa. Even earlier was a handbook on design and construction of microcatchment systems (Shanan and Tadmor, 1976).

In contrast to these manuals, our book is intended to reflect on nearly 30 years of experience, and help those who are looking for pointers to the way forward for water harvesting in Sub-Saharan Africa. An earlier publication by Reij and

(a)   (b)

(c)   (d)

Photo 1.1a–d An example of a 'contour ridge' technology from Kenya: from construction to mature crop (W. Critchley).

colleagues (1996) has acted as a basic model for our approach and format. Its content, too, is closely related. But while those authors sought to establish the credibility of indigenous knowledge regarding soil and water conservation, we take that now as accepted wisdom: one of several changes that have taken place in this field over the last two or three decades. We have also made a deliberate attempt to look for 'bright spots', in the terminology used in Bossio and Geheb (2008), thus drawing attention to successful cases. But equally, we cite examples of less sparkling interventions: some outright failures. Our overall intention is to analyse the experiences with water harvesting and then to propose an agenda for action.

Following two thematically cross-cutting chapters which consider water harvesting in Sub-Saharan Africa as a whole, seven chapters are then dedicated to specific countries. Each of these country-based chapters looks at the history of water harvesting, its context, and its current status and prospects in that country. Figure 1.1 shows the distribution of countries discussed in the book. Then follows a chapter on policy related to water harvesting on the continent and a conclusion.

Chapter 2 sets out the context of water harvesting in the African drylands. Critchley and Scheierling look at the problems of water in agriculture, then trace the history of interventions in sustainable land management in general and water harvesting in particular. The authors move on to give an overview of key studies

Figure 1.1 The distribution of countries discussed in this book.

and assessments since the 1970s, drawing attention to the writings on water harvesting from India and West Asia/North Africa – and pointing out the common denominators as well as the differences. Principles and practices are covered, including a definition of water harvesting in the context of this book: 'the collection and concentration of rainfall runoff, or floodwaters, for plant production'. A classification system is laid out, one that has barely changed since the writings of the 1980s and 1990s. It is stressed that water harvesting needs to be recognized as technically distinct from *in situ* moisture conservation, as the collection of runoff clearly requires different technologies from those that are designed to prevent runoff. Finally the chapter proposes a new concept.

Water harvesting must be seen as more than just technology: thus 'water harvesting plus' or WH+ is proposed. This concept acknowledges aspects of agronomy, fertility management, environmental impacts, livelihoods, incentive mechanisms and enabling policy: issues that are too often overlooked yet have a crucial bearing on the success or failure of water harvesting.

Chapter 3 was conceived by Bouma and colleagues as an analytical update on recent research results through a review of published articles. The focus is on production, livelihoods and uptake. While research results do not always paint the full picture – because there are many factors that determine *what* is researched by *whom*, and furthermore peer-reviewed literature often underplays the views and knowledge of practitioners – several important themes emerge. The first, simply, is that water harvesting literature is on the up: more attention is being paid to the topic. It is no coincidence that conservation agriculture, a rapidly spreading *in situ* conservation technology that also makes judicious use of rainfall, is another subject of increasing research. A second is the confirmation that water harvesting increases yields, and sometimes brings land back into production from a state of degradation. Dramatic improvements are found when water can be stored and used in a controlled way for cash cropping; this is referred to as the 'business case' for water harvesting by the authors. A third is that water harvesting can have significant environmental impacts – usually for the good – including soil conservation, improved biodiversity and increased biomass generally. However the chapter concludes that policy makers are no nearer coming to a general consensus about why there is little uptake of certain systems, though the role of subsidy dependency is clearly one main cause. Finally, and disappointingly, we haven't yet managed to develop a consistent framework for addressing water harvesting impacts.

Chapter 4, covering Burkina Faso, is the first of seven 'country chapters' each of which take a single country where we have both baseline and up-to-date information. Sawadogo and colleagues explain how Burkina Faso was a 'laboratory' for water harvesting in the early 1980s. There were several reasons for this. The first was that the country had experienced severe drought in the 1970s and the government was leading a popular campaign against desertification. Water harvesting was perceived as having an important role to play, and fortuitously there were traditions of building stone lines to slow runoff and digging planting pits (*zaï*) that lent themselves to improvement. These technologies were perfectly suited to reclaiming barren, compacted land. Furthermore, the advent of 'participatory approaches' in the 1980s was taken up with enthusiasm by NGOs and villagers alike, leading to impressive achievements. The chapter tells us about the future evolution of these technologies – the close association with fertility management for example – but we are still unable to find accurate data on the extent of these practices or their impact on livelihoods.

Chapter 5 takes us to Ethiopia, where population pressure combined with recurrent drought and land degradation bring the issue of sustainable intensification of agriculture to the top of the development agenda. Abebe and colleagues note that this is a country where indigenous practices are acknowledged to the extent that the 'Konso cultural landscape' has been recognized as a UNESCO World Heritage Site. Efforts over the last four decades to promote water harvesting have produced some bright spots, particularly with micro-watershed interventions, which include water harvesting as part of an integrated participatory approach to sustainable land management. Other water harvesting technologies examined are floodwater harvesting (usually termed 'spate irrigation' in that country), household ponds and sand dams. Within the chapter there is a case study of the Abreha Watsbeha watershed in Tigray. Here, percolation ponds capturing water from a hillside have contributed to improved groundwater levels, and led to an increase in irrigation but also strikingly visible regeneration of the renowned leguminous 'fertilizer tree' *Faidherbia albida*.

Chapter 6, written by Odour and colleagues, describes interesting evolution in Kenya over the last quarter century. Kenya represented to East Africa what Burkina Faso did to West Africa during those early days of water harvesting projects and trials. Beginning with a detailed history of water harvesting in Turkana District where water harvesting formed the 'work' element of 'food-for-work' the authors show how basic mistakes were made in design and planning by one programme, while simultaneously an NGO-led project helped develop a more rational approach. They look at system development in Baringo District also; here technical success was not matched by spontaneous adoption – and furthermore the invasive exotic, *Prosopis juliflora*, was introduced. The chapter then moves onto a relatively new initiative, driven by farmers themselves: that of road runoff harvesting. This is located mainly in Eastern Kenya, where many farmers now tap runoff from roads and lead it into their farms. One particularly promising model combines runoff capture with ponding of the water and irrigation of vegetables in greenhouses. This is a business model with potential. The authors also show how the 'trapezoidal bund' – an external catchment system from the mid-1980s – has made a comeback and is being promoted once again under 'food-for-assets' (the more creatively named successor to 'food-for-work') in several districts.

Chapter 7 covers Niger. Di Prima, Hassane and Reij look at progress in this regularly drought-prone and food insecure country. Providing three case studies under the Sub-Saharan Water Harvesting Study, Niger was clearly a country of special interest at the time. While the *demi-lunes* microcatchment techniques of Ourihamiza were anticipated to have an uncertain future, they have, in fact, been taken up quite widely by farmers on their own; albeit with modified designs. *Tassa* have proved a great success, and are Niger's equivalent of the *zaï* that have been so popular in Burkina Faso. Their combination with manure and fertilizer has helped many poor farmers to survive. However the SSWHS gave a less rosy prognosis for the mechanized techniques that characterized water harvesting bunds and trenches for tree planting under the Keita Valley project of the Tahou Department. Here it was predicted that these very expensive (and partially mechanized) systems would neither be maintained nor replicated without subsidies. So it has turned out. Finally there is a plea for better monitoring and scientific study, echoing the sentiments of the SSWHS.

Chapter 8 revisits Tanzania where a sustained research and communication effort during the 1990s brought about a remarkable change in perceptions and policy towards rainfall runoff. Whereas previously it had been seen as a hazard which caused erosion and flooding, the possibility of a win–win solution was recognized when it was shown that this runoff could be converted into useful soil moisture storage to improve dryland agriculture. In this chapter, Mahoo and colleagues revisit the original research site and reflect on experience after a gap of 10 years. Faltering successes with the introduction of water harvesting are contrasted with the case of the 'majaruba' rice production system elsewhere in Tanzania which is an undisputed water harvesting bright spot despite receiving almost no external support or promotion.

Chapter 9 highlights the special case of Sudan. Since the SSWHS, Sudan has been divided into two nations, but the study concentrated on the north of the country, which remains Sudan. Critchley, Gaiballa and colleagues point out that with probably the largest extent of water harvesting in Sub-Saharan Africa, Sudan provides both specific and general lessons for the rest of Sub-Saharan Africa. Two technologies are reported on in this chapter: the external catchment system of *teras* from around Kassala and floodwater harvesting/water spreading in the Red Sea Hills. In both cases the originally traditional systems are thriving – in these parts of Sudan there is simply no alternative to making use of water harvesting to grow crops – but there is strong evidence of a trend towards outside support in construction/rehabilitation of structures. Two particularly important issues emerge. The first is that the government has taken a stronger coordinating role of the process of outside help. No longer can agencies 'go-it alone'. The second point raised in the chapter is the concern about the current tendency to increase subsidies for construction of water harvesting, and the possible implications for maintenance and future voluntary construction.

Chapter 10 ends the series of 'country chapters' by taking us to Zimbabwe. Here Gumbo and colleagues explore the legacy of the tied furrow technology studied under the SSWHS and compare its limited impact with the much more promising 'dead-level contour' system. They describe how the particular political situation in Zimbabwe has contributed to stalling progress with water harvesting. The tied furrow system has practically perished for this, amongst other reasons. However the rather oddly named 'dead-level contour' (dead-level implying exactly along the contour) is a system that has attracted a number of followers. As an alternative to the conventional graded ridge (bund) system which drained water out of fields, the dead-level contour holds runoff generated within the farm and from outside also. The chapter is strong on technical detail – but not altogether optimistic about the future of such systems without support agencies being able to operate as freely as they did previously.

Chapter 11 takes up the cross-cutting issue of national policy in regard to water and land resources and other policies that impact on water harvesting. Snelder and colleagues throw the net widely to cover policy regarding the use of water in agriculture and investments associated with this. They note that there has been, in Sub-Saharan Africa, less agricultural water development (especially in rainfed areas) than in any other region in the world. There is also an irony highlighted, that while foreign investment in Sub-Saharan Africa has risen significantly between 1995 and 2010, investment in agricultural water management (from all sources) has been declining. They argue that instruments related to water legislation tend to be discussed with respect to irrigated rather than rainfed farming. Some policy areas where governments can ensure incentives for private sector investments in agricultural water include predictability and reliability of the policy design process; public sector investments are required in associated infrastructure, as well as in training and capacity development. Affordable interest rates on loans must be available to stimulate small-scale farmers to invest. The chapter concludes somberly that unless government policies and investment decisions are supportive, there is a bleak future for small farms in Sub-Saharan Africa.

Chapter 12 than closes the book with a look at what we can glean from the experiences in the foregoing chapters. Why have so many past recommendations remained apparently ignored? This is all the more ironic since the same recommendations continue to be repeated time and again. Have we learnt anything new in the intervening years or merely confirmed what we already suspected? What is the way forward? Gowing and Critchley attempt to answer those questions, and formulate an agenda for action.

Note

1 The project WHaTeR ('Water Harvesting Technologies Revisited: Potentials for Innovations, Improvements and Upscaling in Sub-Saharan Africa') is funded through the European Commission's FP7 Instrument. A consortium of eight partners from an equal number of different countries are involved. The project began in 2011 with a systematic programme of revisits to countries where water harvesting had been documented 20–25 years ago. We are grateful to the project for helping provide much of the information that this book contains.

References

Bossio, D. and Geheb, K. (eds) (2008) *Conserving Land, Protecting Water*, CABI International, Wallingford, UK.

Critchley, W. R. S. and Siegert, K. (1991) *Water Harvesting: A Manual for the Design and Construction of Water Harvesting Schemes for Plant Production*, FAO, Rome.

Critchley, W. R. S., Reij, C. and Seznec, A. (1992) 'Water harvesting for plant production, Volume II: Case studies and conclusions for Sub-Saharan Africa', World Bank Technical Paper Number 157, Africa Technical Department Series, Washington D.C.

Dutt, G. R., Hutchinson, C. F. and Garduno, M. A. (eds) (1981) *Rainfall Collection for Agriculture in Arid and Semi-Arid Regions*. Proceedings of a Workshop, University of Arizona and Chapino Postgraduate College, Commonwealth Agricultural Bureaux, UK.

Evanari, M., Shanan, L. and Tadmor, N. H. (1971) *The Negev – The Challenge of a Desert*, Harvard University Press, Cambridge, MA.

Fallon, L. E. (1963) *Water Spreading in Turkana: A Hope for an Impoverished People*, Mimeo, USAID, Nairobi, Kenya.

Nissen-Pietersen, E. (1982) *Rain Catchment and Water Supply in Rural Africa: A Manual*, Hodder and Stoughton, London.

Oweis, T. Y., Prinz, D. and Hachum, A. Y. (2012) *Rainwater Harvesting for Agriculture in Dry Areas*, CRC Press, London, New York, Leiden.

Pacey, A. and Cullis, A. (1986) *Rainwater Harvesting: the Collection of Rainfall and Runoff in Rural Areas*, IT Publications, London.

Reij, C., Mulder, P. and Begemann, L. (1988) 'Water harvesting for plant production', World Bank Technical Paper 91. Washington D.C.

Reij, C., Scoones, I. and Toulmin, C. (1996) *Sustaining the Soil: Indigenous Soil and Water Conservation in Africa*, Earthscan, London.

Shanan, L. and Tadmor, N. H. (1976) *Micro-catchment systems for Arid Zone Development. A Handbook for Design and Construction*, Ministry of Agriculture, Centre for International Agricultural Cooperation, Rehovot, Israel.

Steenbergen, F., Lawrence, P. and Haile, A. M. (2010) 'Guidelines on spate irrigation', FAO Irrigation and Drainage Paper 65, Rome.

Chapter 2

Water harvesting for crop production in Sub-Saharan Africa

Challenges, concepts and practices

William Critchley and Susanne M. Scheierling

Introduction

This chapter sets out the position and importance of water harvesting within water-constrained rainfed agriculture. After reviewing the challenges related to rainfed agriculture, it touches on history and evolution of the key concepts and practices of water harvesting in Sub-Saharan Africa. This is followed by sections on the classification of water harvesting, a comparison of the key technologies, and a closer look at common practices in Sub-Saharan Africa. Finally, the chapter discusses a new approach to more effective water harvesting.[1]

Rainfed agriculture: challenges, and the role of water harvesting

In water-constrained rainfed production systems, rainfall-driven (or, more accurately, climate-driven) variability leads to low and unstable production and productivity, and is often the dominant source of income and consumption risk for farmers (Dercon, 2002; Zimmermann and Carter, 2003). Climate-driven fluctuations in agricultural production contribute substantially to the volatility of food prices, particularly where remoteness, the nature of the commodity, transportation infrastructure, stage of market development, or policy, limit integration with regional or global markets. Because market forces tend to move prices in the opposite direction to the level of crop production, increases in food crop prices will buffer farm incomes, but exacerbate food insecurity for poor consumers.

Several relationships that determine the productive potential of rainfall during the growing season of a crop need to be kept in mind when discussing possible interventions for improving water management in rainfed agriculture. If rainfall is less than crop water requirement, then actual yield will be less than the potential. Equally important is the distribution of rainfall. If rainfall fulfils 70 per cent of crop water requirements *every day*, then a good yield is possible, but if rainfall is 100 per cent of crop requirements for 70 per cent of the growing season and zero for the rest, the outcome will be quite different. Moreover, the impact of variable rainfall is strongly affected by the characteristics of the soil and the stage of the growing season. If the soil is capable of storing a large quantity of water in relation to crop demand, then a break in rainfall of a week or more may be tolerable, especially late in the season. Conversely, in a semi-arid climate where daily demand is high and the soil is less capable of storing water, crop sensitivity to breaks in the rainfall is very high.

An additional complication is that, while biomass varies directly with evapotranspiration (Howell, 1990; Perry et al., 2009), the proportion of grain or fruit in the total biomass (the harvest index) is sensitive to when water stress occurs (Fereres and Soriano, 2007). A gap in the rainfall at a critical stage in crop development (e.g. at flowering and grain-fill stages) will be far more damaging to eventual yield than a break at a less sensitive stage.

Thus, while all rainfed farmers face risks due to rainfall variability, the risk increases when the total amount of rainfall is low in relation to crop water requirements; when rainfall occurs erratically and in a few events that contribute largely to runoff; when long gaps between rainfall events are frequent; when soils are not retentive; and when daily crop water requirements are so high that soil moisture storage is rapidly depleted. Farmers in a risky environment need to balance the cost of adding extra inputs against the potential benefits in terms of increased crop value, taking into account the likelihood that the rainfall pattern may be poor. *Ex post*, once the actual rainfall pattern is known, it is not difficult to determine the optimum farming strategy, but the problem arising for farmers is to make decisions *ex ante*.

To exacerbate the problem of water management in rainfed areas of Sub-Saharan Africa, land degradation is not only serious (Bridges et al., 2001), but according to new research is getting worse in many areas (Vlek et al., 2008). Land degradation impoverishes the resource base, interferes with ecosystem function, releases carbon to the atmosphere and threatens livelihoods. Increasing soil nutrient deficiencies is also a problem (Figure 2.1), as water management without fertility improvement is frequently not sufficient for significant production increases. In Sub-Saharan Africa the majority of farmland is rainfed: less than 5 per cent is currently irrigated. Climate change will bring new problems, with drying expected in the Sahel and Southern Africa, and though East Africa may become wetter this is at the expense of more intense and erosive rainfall events (IPCC 2007; Toulmin, 2009). Rockström and colleagues have led the call to recognize – and exploit – the untapped potential of rainfall in Sub-Saharan Africa. Rockström (2000) estimates that up to 70–85 per cent of rainfall is 'lost' in the Sub-Saharan drylands. Furthermore Rockström and Falkenmark (2000) point to the possibility of 'doubling crop yields with small manipulations' of rainwater (p. 319). If rainfall was better translated into 'productive green water flow' (i.e. water that is directly used for plant growth) rather than lost through runoff, surface evaporation or deep drainage, then especially at low levels of productivity, yields and water use efficiency could be improved dramatically. Molden (2007) in the benchmark report on the *Comprehensive Assessment of*

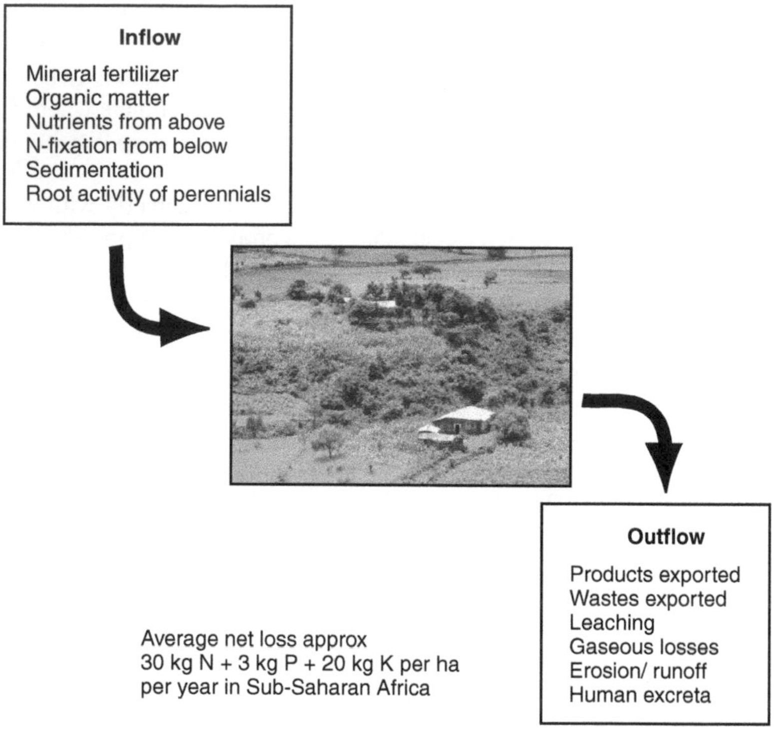

Inflow

Mineral fertilizer
Organic matter
Nutrients from above
N-fixation from below
Sedimentation
Root activity of perennials

Outflow

Products exported
Wastes exported
Leaching
Gaseous losses
Erosion/ runoff
Human excreta

Average net loss approx
30 kg N + 3 kg P + 20 kg K per ha
per year in Sub-Saharan Africa

Figure 2.1 Nutrient flux in small-scale farming in Sub-Saharan Africa (Critchley 2010: after Hilhorst and Muchena, 2000)

Water Management in Agriculture states that: 'The greatest potential increases in yield are in rainfed areas where many of the world's poor live and where managing water is the key to such increases' (p. 2).

Globally, rainfed farming has to be the key player in achieving the goal of feeding the projected nine million people by 2050 (Koohafkan and Stewart, 2008). 'Sustained intensification' is the new clarion call for improving productivity without damaging the environment. Irrigation has largely reached its limits in many areas, and the next green revolution will need to target rainfed areas. Wani et al. (2009) come to the conclusion that there is very considerable untapped potential for rainfed agriculture – and investment in better water management is crucial in this context. In their 'optimistic' yield-growth scenario, which assumes that 80 per cent of the current gap between actual and obtainable yields would be bridged, 85 per cent of projected food demand by 2050 could be met by rainfed farming with an expansion of only seven per cent of rainfed land. This points to the importance of improving systems of rainfed agriculture that make optimum use of rainfall and are embedded in sustainable land management. Two of the practices with the most potential are conservation agriculture (an *in situ* water conservation technology based on no-till, mulching and crop rotation) in the more humid areas, and water harvesting in the drier zones. To achieve these goals, it is imperative that existing (but often neglected) recommendations – both in terms of technologies and approaches – are revived and acted upon alongside new research. The exigencies of climate change and population growth leave few other options.

History of water harvesting

Trends and milestones

Before turning to water harvesting in more detail, it is well to explore the history of concern about rainfed agriculture, water stress, soil erosion, desertification and remedies. Modern international consternation about these issues can be traced back to the 1930s with the 'Great Dust Bowl' phenomenon in the United States, which resulted in the first soil conservation campaigns being carried out. While water harvesting itself has roots in ancient tradition, it gained wider prominence during the African droughts of the 1970s. However, many failed water harvesting and soil conservation schemes led to reappraisal and significant shifts in conceptual thinking. The evolution of approaches in rainfed agriculture and water harvesting is traced in Table 2.1 which summarizes trends and milestones.

From the 1980s on, it can be said that a new overall approach emerged. This approach is broadly characterized (in rhetoric, at least) by participatory, production-oriented, affordable strategies of conservation (WOCAT, 2007). This more 'enlightened' attitude is held to be more likely to spread success, particularly amongst resource-poor farmers (Chambers, 1983; Douglas, 1994; Hudson, 1991; IFAD, 1992). Terms and concepts have moved on also, keeping pace with the new thinking. Thus 'soil conservation' progressed first to 'soil and water conservation' or 'soil and water management'. Then 'better land husbandry' was popularized (Shaxson et al., 1989), followed by 'sustainable land management' which consolidated its position in 1996 at the International Soil Conservation Organisation (ISCO) conference in Bonn (Hurni et al., 1996) and reflected the new international focus on 'sustainable development' (WCED, 1987). 'Sustainable land management', with its central notion of combining conservation with production, continues to be the preferred term in the anglophone world. Recent years have seen the re-emergence of views of sustainable agriculture as the 'engine of growth' in poor countries. This resonates well with the realization that in many sub-Saharan countries dryland farming communities are among the poorest of the poor while also being the main custodians of the countryside.

In recent years the climate change community has also acknowledged that inappropriate agricultural practices and land use change are major sources of greenhouse gases (IPCC, 2007). While most climate change data is open to

Table 2.1 Trends and milestones: Rainfed agriculture and water harvesting in Sub-Saharan Africa

Dust Bowl trigger for start of international conservation schemes	1930s	Great Dust Bowl in the USA: Soil Conservation Service set up
	1940s	Coercive terracing programmes in a number of African colonies
Terracing promoted strongly	1960s	Independence of many African countries: soil and water conservation activities dropped
	1972	Stockholm conference on the environment; establishment of UNEP
Terracing rejected as 'colonial'	1970s	Soil conservation/rural development projects fail in Sub-Saharan Africa and elsewhere
Soil conservation projects & many fail	1974	*More Water for Arid Lands* (National Academy of Sciences, on Water Harvesting)
Agroforestry 'discovered'	1977	Nairobi Conference on Desertification
Analytical lessons drawn from new wave of projects	1977	'Agroforestry' named as a science; World Agroforestry Centre (ICRAF) set up in 1978
	1982	*Review of Rainwater Harvesting* (Boers and Ben-Asher, on Israel, USA, among others)
Participatory approaches introduced	1983	*Rural Development: Putting the Last First* (Chambers, on participatory methods)
	1986	*Rainwater Harvesting* (Pacey and Cullis, on water harvesting in Sub-Saharan Africa and elsewhere)
Water harvesting becomes subject of study and project implementation	1986	Applied geography 'theme' volume on runoff farming in rural arid area (Briggs, 1986)
	1986	*Lessons of Water Harvesting from Sub-Saharan Africa* in Baringo, Kenya (Critchley, 1987)
	1987	*Our Common Future* (WCED's Brundtland Report on sustainable development)
Indigenous practices gain credence	1989	*La Lutte Contre la Desertification* (Rochette, on experiences in the Sahel)
Sustainable land management accepted as new term	1991	FAO's *Water Harvesting* (Critchley and Siegert)
	1992	Rio Conference
	1992	World Bank's *Water Harvesting for Plant Production* (Critchley et al.)

Table 2.1 Cont'd

Water harvesting/ Indigenous knowledge awareness in India	1993	FAO's *Water Harvesting for Improved Agricultural Production* (workshop and book)
	1995	'Blue water', 'green water' concept introduced by Falkenmark
	1997	*Dying Wisdom* (Agarwal and Narain, on traditional water harvesting in India)
Conservation agriculture takes off in the Americas and elsewhere	2000	Conservation agriculture begins to expand rapidly
	2001	*Water Harvesting for Upgrading of Rainfed Agriculture* (Falkenmark et al.)
Global issues take centre stage	2001	*Making Water Everyone's Business* (Agarwal et al., on water harvesting in India)
Agriculture (especially small-scale and rainfed) re-emphasized	2004	*Indigenous Water-Harvesting Systems in West Asia and North Africa* (Oweis et al.)
	2007	WOCAT's *Where the Land is Greener: Case studies of SWC Worldwide*
	2007	*Comprehensive Assessment of Water Management in Agriculture* (Molden, ed.)
Rainfed farming and Water harvesting regain attention	2008	World Bank's World Development Report *Agriculture for Development*
	2009	*Rainfed Agriculture: Unlocking the potential* (Wani et al.)
	2010	FAO's *Guidelines on Spate Irrigation* (Steenbergen et al.)
	2011	FAO/TerrAfrica's *Sustainable Land Management in Practice* (Liniger et al.)
	2012	*Rainwater Harvesting for Agriculture in Dry Regions* (Oweis et al.)

question with respect to precision, there is no doubt that very considerable amounts of greenhouse gases, around 14 per cent of the total worldwide, are derived from agriculture. These include methane from farm animals and rice paddies, nitrous oxide from nitrogen fertilizers, and carbon dioxide from farm machinery. An additional 17 per cent of greenhouse gases are from deforestation, mainly carbon dioxide from decaying vegetation, and from exposure of soil organic matter (IPCC, 2007; World Bank, 2011). Sustainable land management is thus not simply a local environmental matter: it also central to concerns about poverty and climate change. This is also the basic tenet of 'climate-smart agriculture' with its triple wins of food security, resilience and increased carbon sequestration (World Bank, 2011).

Key studies and assessments

To substantiate the historical timeline outlined in Table 2.1, it is worth briefly reviewing some publications that have influenced the thinking on water and soil management in rainfed agriculture within dry regions. Several of these writings (those pertaining to water harvesting in particular) are included in this chapter with a focus on the period since the mid-1970s. Chapter 3 takes this review further by discussing in greater detail the publications from the last two decades. The reason for going back several decades is that important older studies, reviews and assessments are in danger of 'disappearing'. This is partially because many have not been digitized (and thus web-based searches will overlook them or not provide access to content), partially because there is a tendency to ignore what has been done before (with field research often more appealing than an analysis of insights from the literature), and partially because a large amount of information is currently being generated that tends to 'bury' the older material.

A seminal publication was *More Water for Arid Lands* by the National Academy of Sciences of the United States which drew attention to 'runoff agriculture' (NAS, 1974). It was followed by an overview of water harvesting by Boers and Ben-Asher (1982). With regard to water harvesting in Sub-Saharan Africa, a publication by Pacey and Cullis, part review and part design manual, was innovative and pivotal (Pacey and Cullis, 1986). A series of droughts in Africa had drawn attention to the plight, but also the potential, of semi-arid areas. One of the main conclusions of Pacey and Cullis was that '[i]nformation about existing traditions of runoff farming is inadequate nearly everywhere' (p. 127).

During the period of heightened project activity in the 1980s, following on from the severe Sahelian droughts of the 1970s, Rochette travelled extensively around francophone West Africa, visiting projects in Burkina Faso, Mali, Mauritania, Niger and Senegal. His ensuing book focused on the experience of 21 projects, giving a broad coverage of countries, agro-ecological zones and technologies (Rochette, 1989). Each experience is treated systematically, discussing the historical context and the details of the project's technical activities and socio-economic impacts, and providing conclusions. While a major work, because it was only available in French it had little impact on anglophone Africa. This linguistic divide, a legacy of colonialism, has continuously hampered exchange of experience.

Also during the 1980s, the United Nations Food and Agriculture Organization (FAO) identified the need for guidance on water harvesting, and commissioned a technical handbook. The objective was to provide technicians and extension workers with practical guidelines on the implementation of water harvesting schemes. The resulting handbook (Critchley and Siegert, 1991) provides a classification of water harvesting systems, covers water and soil requirements for water harvesting, and presents an analysis of rainfall and runoff as an important input for designing water-harvesting interventions. The main section of the handbook is dedicated to discussing the technical details of various systems, covering the most common project-based technologies from East and West Africa.

Much of the information provided in the handbook was derived (with permission) from the World Bank's Sub-Saharan Water Harvesting Study (reported in Critchley et al., 1992). The latter study marked a renewed interest in water harvesting by the World Bank, and was stimulated by positive results from water harvesting under the Baringo Pilot Semi-Arid Area Project in Kenya (MoALD, 1984) and a later workshop that highlighted the project's activities and the general interest in the topic within Sub-Saharan Africa (Critchley, 1987). Water harvesting components were characteristic of a number of projects across Africa at the time, and there was a growing realization that indigenous systems were largely unstudied – or even unknown. Based on field visits and a literature review, Critchley et al. (1992) describe 13 case studies from seven different countries and highlight the importance of not just technologies/structures that form the basis for all water harvesting systems, but also production aspects (such as crops, cropping systems, and the need for fertility maintenance). Under socio-economic and project management aspects, issues such as adoption, mechanization, land tenure, participation, costs and benefits, and the use of incentives are discussed. Besides various technical points, the recommendations of the report included the need for coordination at national level; the requirement for research into non-adoption; the importance of land tenure; the need to recognize and study traditional systems; and the urgency of setting up monitoring and evaluation systems. While not all recommendations were original, none of them are outmoded, and many have been reiterated and further developed in subsequent publications over the years.

With respect to India, where water harvesting is also important, especially as a means of replenishing groundwater, Kerr and Sanghi (1992) set the ball rolling by their review of traditional systems of soil and water conservation. Five years later, *Dying Wisdom* represented India's response to the recognition of ancient water harvesting systems and their potential for the future (Agarwal and Narain, 1997). The book is comprehensively researched and well illustrated and, while not a design manual, it is a readable and impressive overview of systems from India's 15 ecological zones. Both agricultural and domestic water supply systems of water harvesting are covered. The book was followed up in 2001 by the more analytical, and geographically broader, *Making Water Everybody's Business* (Agarwal et al., 2001).

The International Centre for Agricultural Research in Dry Areas (ICARDA) launched a regional initiative on 'on-farm water husbandry in West Asia and North Africa' in 1996. The resultant publication is a compilation of case studies of indigenous systems from Egypt, Iraq, Jordan, Libya, Morocco, Pakistan, Syria, Tunisia and Yemen (Oweis et al., 2004). The editorial chapter stresses the role of indigenous knowledge (and local innovation) in improving existing systems through blending with western 'scientific' knowledge. The history of water harvesting in this region – the cradle of agriculture – is as old as anywhere in the world.

The publication reports on some of the most ancient systems that still prosper, and a range of newer technologies that are being introduced. A follow-up overview and comprehensive technical design manual has recently been published (Oweis et al., 2012).

At a global scale, the World Overview of Conservation Approaches and Technologies (WOCAT) has, since the early 1990s, been systematically building up a database both of technologies (including soil conservation, sustainable land management, and water harvesting) and the 'approaches' (the enabling environment) that support those technologies. *Where the Land is Greener* (WOCAT, 2007) was the programme's first global publication. Forty-two technologies are reported as case studies – three of which are related to water harvesting (namely planting pits and stone lines from Niger, streambed structures from India, and microcatchments for olives from Syria) – and the majority are matched by associated 'approaches'. A key point made by the editors is that quantitative data on the various impacts of the technologies was hard to find. Among the key policy points emanating from the book's analysis are the following: monitoring and evaluation in projects or programmes must be improved; mapping of conservation coverage is essential; further research is needed on various technical and social parameters; in dry areas, investments in water harvesting and improved water use efficiency, combined with improved soil fertility management should be emphasized; and local innovation and farmer-to-farmer exchange should be promoted. Another publication by WOCAT in conjunction with TerrAfrica focuses on systems specific to Africa and provides a series of new case studies, including a section on water harvesting with examples from Ethiopia (floodwater harvesting), Niger (planting pits) and Zambia (small earth dams) (Liniger et al., 2011).

The *Comprehensive Assessment of Water Management in Agriculture* by Molden et al. (2007) sets out the prospects for water in both rainfed and irrigated agriculture. Not surprisingly there is a note of warning, yet guarded optimism, in its key quote: 'Only if we act to improve water use in agriculture will we meet the acute freshwater challenges facing humankind over the coming 50 years' (p. 1). Of the eight policy actions proposed, the one most relevant here is the following:

> Upgrade rainfed systems – a little water can go a long way. Rainfed agriculture is upgraded by improving soil moisture conservation and, where feasible, providing supplemental irrigation. These technologies hold underexploited potential for quickly lifting the greatest number of people out of poverty and for increasing water productivity, especially in Sub-Saharan Africa and parts of Asia. (p. 4).

A series of publications surround this central study, one of them being *Rainfed Agriculture: Unlocking the Potential* (Wani et al., 2009). Many of the constraints discussed in this publication will sound familiar to those conversant with the related literature over the years. These include lack of a market-oriented smallholder production system, poor research–extension–farmer linkages, the need to focus on soil fertility improvement, and the urgency of strengthening capacities of institutions and farmers' organizations. The strategies put forward are broadly those that address these constraints. Amongst the technical remedies proposed is water harvesting which is discussed in two chapters, one with a focus on the Middle East and North Africa, and the other principally on India.

Concepts and practices

Concepts of water harvesting

Water harvesting basically aims to augment limited rainfall on a target area through harnessing runoff or floodwaters. In cases where there is not enough rainfall to allow adequate plant growth in a particular area, then it is logical to concentrate it in one location and forgo agricultural production elsewhere. Water harvesting is thus a deliberate reallocation of a resource within the landscape in order to improve overall agricultural production, by manipulating a natural process. Figure 2.2 illustrates this for a microcatchment system where a cross section through a contour ridge, examined after a rainfall event of 53mm, shows that the resulting runoff almost doubles the effective rainfall in the cropped area.

Farmers usually employ one of three basic strategies to manage water under rainfed conditions (Narayana and Babu, 1985, quoted in Hudson, 1987). Depending on the amount of rainfall, potential evapotranspiration, and the cropping system, farmers may choose systems or technologies that *primarily*: (i) provide for discharge of excess water, (ii) hold rainfall *in situ*, or (iii) harvest water from a catchment area to supplement rainfall.

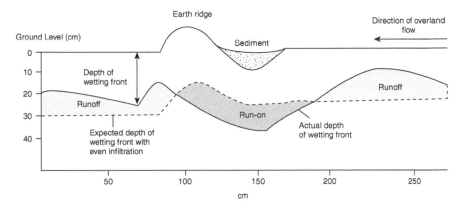

Figure 2.2 Cross section of contour ridge after rainfall event: showing expected wetting front with even infiltration, and actual wetting front because of water harvesting effect (based on MoALD, 1984)

There is a continuous variation through these systems, and thus no precise cut-off points. But the three categories help to distinguish primary strategies in different situations, and to avoid confusion in terminology. Oweis et al. (2012) clarify the difference between the second and the third category: 'Soil-water conservation practices aim at preventing surface runoff and keeping rainwater in place, whereas water harvesting makes us of, and even induces, surface runoff' (p.3). Thus conservation agriculture, for example, is not a water harvesting technology: it functions through conserving rain where it falls.

As described below, water harvesting systems vary in many ways. One of the most important is the scale, and the ratio between catchment and cultivated area (C:CA). Thus there are floodwater harvesting systems where the catchment is hundreds of times the size of the cultivated area and, on the other hand, microcatchment systems where the ratio between catchment and cultivated area is as little as 1:1. A second key difference is the method of water storage. Harvested runoff may be stored in the soil profile, or it may be ponded. The range of technologies associated with different water harvesting systems is enormous. It would be wider still if other uses of harvested water beyond plant production were to be included, such as water supply for domestic purposes and livestock.

Definition and classification

For the purposes of simplicity and consistency, the definition for water harvesting used here is based on that proposed by Critchley and Siegert (1991), namely 'the collection of runoff for its productive use' (p. 4). When further refined to include floodwater harvesting and explicitly adding the element of water concentration, as well as this chapter's focus on plant production, water harvesting is defined here, and in the context of this book, as: 'The collection and concentration of rainfall runoff, or floodwaters, for plant production.'

The basic components of a water harvesting system that is dedicated to plant production are (i) a catchment area, (ii) a concentration area, and (iii) a cultivated area. When runoff is stored in the soil profile, the concentration area and the cultivated area are synonymous.

Figure 2.3 presents a classification of water harvesting systems simplified from that developed by Critchley and Siegert (1991) which was repeated in FAO (1994). The classifications proposed by Pacey and Cullis (1986), Oweis et al. (1999), Fox (in Falkenmark et al., 2001), Oweis et al. (2004) and Oweis and Hachum (2009) differ in some details, but are variations on the same theme. The classification in Figure 2.3 divides water harvesting into floodwater harvesting (channel flow) and rainwater harvesting (overland flow), and differentiates between microcatchments, external catchment systems and floodwater harvesting. The final subdivision is between systems that store runoff in the soil and those that pond water for supplementary irrigation. Storing rainfall runoff in the soil can be achieved by ridges or bunds, and does not require lifting/pumping equipment.

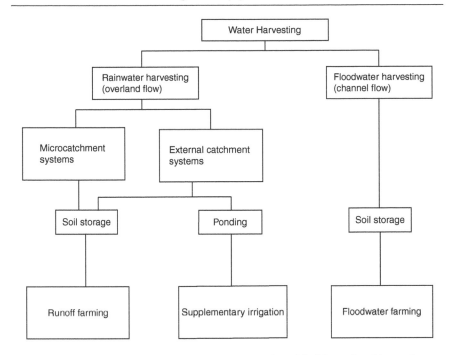

Figure 2.3 Classification of water harvesting systems (simplified from Critchley and Siegert, 1991)

However, because the crop is dependent on that limited storage, these systems may become vulnerable to prolonged drought: water harvesting only magnifies the impact of rainfall. Systems that pond water tend to be less vulnerable (see Chapter 3). But this comes, literally, at a cost.

It should be noted that floodwater harvesting, water spreading and spate irrigation are related terms that are often used interchangeably. While floodwater harvesting and water spreading are broadly synonymous, they do not quite coincide with spate irrigation. Spate irrigation definitions *include* much floodwater harvesting/water spreading, but spate irrigation schemes generally have better developed infrastructure, and some of these come close to conventional irrigation where the water supply is more secure with significant base flows (see van Steenbergen et al., 2010). The term preferred here, and throughout this book, is floodwater harvesting; though in some countries, for example Ethiopia, spate irrigation is used as a generic term, and in others water spreading is more common.

Table 2.2 compares and contrasts the characteristics of microcatchment, external catchment and floodwater harvesting/water spreading systems. Microcatchments make use of local surface runoff/overland flow, often from within the field itself. They are characterized by a high runoff coefficient, and crop growth is generally even. But they may be ineffective without several runoff

Table 2.2 A comparison of microcatchment, external catchment and floodwater harvesting systems

	Rainwater harvesting		Floodwater harvesting
Characteristics	Microcatchment	External catchment	
Location of catchment	Within field	Outside field: hillside/ road etc.	Large catchment: often hilly terrain
Length of catchment	A few metres	Often 50 metres or more	Several kilometres
Type of flow	Overland flow	Overland flow (sometimes rills or small channels)	Channel flow
Landscape	Even over whole area	Landscape divided up into catchments and plots	Clear distinction between hilly catchment zone and cropped fields on plain
Catchment: Cultivated Area ratio (C:CA)	Low – usually from 1:1 to 5:1	Medium – usually from 3:1 to 20:1	High – can be 100:1 or more
Runoff coefficient	Relatively high	Relatively low: the longer the catchment the lower the coefficient	Relatively low
Storage of runoff	In strips/pits in soil within-field	In soil over whole of field, or ponded	In soil over whole planted area
Spillway	Not required	Needed: to discharge excess runoff	Essential: usually bunds acts as baffles, deflecting and slowing the flow
Type of crop	Most crops	Crops tolerant of temporary waterlogging or rapidly maturing on residual moisture/high value crops when water stored in ponds	Crops tolerant of temporary waterlogging or rapidly maturing on residual moisture
Planting configuration	In strips/pits within-field	Evenly over whole of field	Evenly over the whole planted area
Fertility	No sediment delivered	Some sediment/animal droppings captured	Often 'self-fertilizing' with sediment
Suitability	Especially where rainfall is reliable	Especially where few runoff events expected	Driest areas with ephemeral watercourses
Social organization	Usually individual	Usually individual	Requires community organization

events per year. External catchment systems also harvest overland flow, plus some sediment and organic litter, from outside the cropped area – for example from a hillside or road. They can function well with few runoff events, and thus are a good option for drier areas. However, design requires careful attention to safe disposal of surplus runoff through spillways. Floodwater harvesting captures channel flow, typically from ephemeral watercourses. The flow often carries considerable sediment which can fertilize the cropped area. These systems are suited to the driest zones, and generally require a higher degree of social organization than the other forms of water harvesting as they function on a community basis.

Table 2.3 lists the main water harvesting technologies that are discussed in the following chapters of the book, under the categories given above. Twelve practices are described, though some cover more than one technology where they are closely related: for example, the closely related 'dead-level contours' from Zimbabwe and infiltration ditches from Uganda.

'Water harvesting plus': a new concept

The science of water harvesting is not solely concerned with the technology of capturing runoff and satisfying crop water requirements. However, texts on the subject have often focused on this specific aspect, and engineering design based on hydrological considerations has been in the driving seat. Much research has been devoted to runoff coefficients, peak flows, and crop water requirements in relation to rainfall probabilities – all factors that determine important structural dimensions such as the ratio of catchment to cultivated areas. While these aspects are crucial, water harvesting is much more than that. It can be considered a specific form of 'sustainable land management' which is also multifaceted (Hudson, 1981; WOCAT, 2007). As noted above, it is instructive that the concept of sustainable land management has evolved over the past decades from 'soil conservation', signalling an increasingly holistic approach. In theory at least, it now incorporates notions of engineering and agronomic interventions while embracing climate adaptation and mitigation, ecosystem thinking, as well as socio-economic and policy aspects. During the same period, from the 1970s to present, 'water harvesting' has remained static in name, and it can be argued that approaches to the discipline have in many cases not significantly evolved past engineering concerns (an exception is, for example, Scheierling et al., 2012).

However, to be effective, water harvesting must also link the elements of efficient water utilization with sound agronomic practices; thus fertility management (through organic manuring, incorporation of legumes and micro-dosing with fertilizers), method of tillage, weed control, choice of crop and variety (drought tolerant or drought evasive) and agroforestry mixtures are all crucial. In the case of floodwater harvesting systems, these aspects also apply, with the exception of fertility management. Floodwater harvesting tends to be self-fertilizing; that is, it

Table 2.3 Common water harvesting technologies in Sub-Saharan Africa. Images based on (from top to bottom): Borst and de Haas (2006), Critchley (1991), Critchley et al. (1992), Critchley (1999), Critchley and Siegert (1991), Finkel et al. (1987), MoALD (1984)

Technology Local name	Classification	Description	Where found in SSA Relevant Chapters	
Planting pits Zaï Tassa	Microcatchment	Deep (about 20 cm), wide-spaced (about 100 cm) planting pits with several plants in each. Earth piled on lower side. Manure applied. Captures runoff within field.	Particularly in West African Sahel – but some examples found elsewhere. Chapter 4: Burkina Faso Chapter 7: Niger	
Semi-circular bunds Demi-lunes and Negarim	Microcatchment	Semi-circular shaped earth bunds with tips on contour, usually in a staggered formation. Crops planted evenly within each structure. Negarim similar but V-shaped with tree in apex.	Widespread. Chapter 4: Burkina Faso Chapter 6: Kenya Chapter 7: Niger	
Contour ridges/bunds	Microcatchment	Earth ridges (about 20 cm high) at 1.5–3.0 m apart on contour, or larger bunds (about 50 cm high) at 5–10 m apart, collecting runoff between structures.	Reported from Kenya but variations found elsewhere also. Chapter 6: Kenya	
Tied furrows	Microcatchment	Broad low ridges with wide furrows with earth ties to prevent lateral flow. Catchment is the ridge and concentration area the furrow.	Reported from Zimbabwe but variations found elsewhere also. Chapter 10: Zimbabwe	
Stone lines/bunds Diguettes en pierres	External catchment	Low stone bunds on the contour at a spacing of 15–30 m apart slowing and filtering runoff from external catchment.	Particularly in West African Sahel – but some examples found elsewhere. Chapter 4: Burkina Faso	
Trapezoidal bunds	External catchment	Trapezoidal shaped (from above) earth bunds capturing runoff from external catchment and over-flowing around stone-reinforced wingtips.	Kenya Chapter 6: Kenya	

Technology Local name	Classification	Description	Where found in SSA Relevant Chapters	
Teras	External catchment	Earth bunds on three sides of a cultivated plot with upslope side open to capture runoff from catchment. Mini-terasmay be established within main teras.	Sudan

Chapter 9: Sudan | |
| Dead-level contours/ Infiltration ditches | External catchment | Level contour bunds with deep furrows upslope capturing runoff from outside and within the field. Covered infiltration pits set within the furrows. | DLC specific to Zimbabwe though infiltration ditches common throughout SSA

Chapter 6: Kenya
Chapter 11: Zimbabwe | |
| Farm ponds | External catchment | Ponds usually ranging from 30–1,000m^3 in size capturing runoff water from an external catchment (such as hillside or road). Water extracted for supplementary irrigation. | Common throughout SSA

Chapter 5: Ethiopia
Chapter 6: Kenya | |
| Sand dams | Floodwater harvesting | Impermeable barriers constructed across ephemeral/seasonal rivers to harvest water and store in sand. Water extracted through wells dug in bank. For domestic use as well as supplementary irrigation. | Almost exclusively found in East Africa.

Chapter 5: Ethiopia
Chapter 6: Kenya | |
| Permeable rock dams Digue filtrante | Floodwater harvesting | Long (up to 100 m), low but broad rock dams across valley floors, slowing and spreading floodwater as well as rehabilitating the gullies. | West African Sahel.

Chapter 4: Burkina Faso
Chapter 7: Niger | |
| Floodwater harvesting/ Water spreading/ Spate Irrigation | Floodwater harvesting | Various configurations based on diverting floodwater from an ephemeral watercourse. Diversion may be concrete, gabions, stone or earth. Bunds used for in-field spreading. | Widespread over SSA though most common in East and West.

Chapter 8: Tanzania
Chapter 9: Sudan | |

annually replenishes the flooded fields with silt. At the smallest scale, 'gully gardens' are a perfect example of this: the stone (or wood) barrier across the gully is specifically designed to capture sediment as well as water. Critchley et al. (2008) describe this as deliberately 'hijacking the water for its bounty of sediment' (p.115). However, in other systems of water harvesting, improving water availability to plants may unmask soil fertility as the second limiting factor, and this has to be carefully managed (see Chapter 3). Figure 2.4 shows not only how water harvesting can significantly raise cereal yields at the lower margins of water availability, but also how the response can be even more pronounced with good agronomic management. While the figure presents simplified, aggregate values (some of the complexities of these relationships have been discussed above), it demonstrates the principle of water harvesting and the potential to enhance yield impact with good agronomic practices.

Water harvesting is also a strong candidate to be acknowledged as a climate change adaptation intervention (see, e.g. Ngigi, 2009). This is for two reasons. The first is that it helps address drought and dry periods in the cropping season,

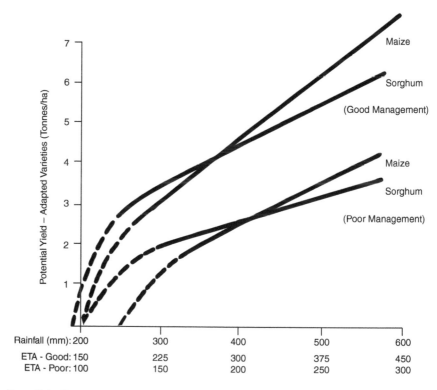

Figure 2.4 Cereal yield, rainfall and management practices showing predicted crop response to rainfall under different management regimes (after Whiteman, 1981)

phenomena that are likely to become more important with a changing climate. The second is that water harvesting systems can themselves be adapted by land users. Catchment size can be manipulated by diverting flows towards or away from the cropped area. But there are also possibilities of utilizing husbandry skills of opportunistic 'response farming'. This includes relay planting of a legume, for example, into a standing crop if climatic conditions are favourable, or making use of an out-of-season flood event to seed a catch-crop (a rapidly maturing plant) to utilize moisture stored in the soil. It is relevant in this context to reiterate that water harvesting is an ancient tradition, one that has constantly been adapted by land users, through their innovation, as a response to changing conditions.

Furthermore, ecosystem thinking is important under water harvesting (Barron, 2009). The capture of potentially erosive runoff helps prevent land degradation, and that in turn has a positive influence on ecosystems and associated biodiversity if water harvesting is practiced at scale. But there is a caveat: for example, some flood-water harvesting systems may divert significant flows away from natural ecosystems located lower in the catchment, to the latter's detriment (see Chapter 9 on Sudan).

Finally, water harvesting involves numerous non-technical aspects also. For example, water harvesting has had a complex and often uncomfortable relation-ship with drought relief since the first series of schemes were introduced in Sub-Saharan Africa during the 1980s. The use of incentives and subsidies – including food-for-work – has been common both during emergencies and after-wards (see Chapters 5, 6 and 7). Such policies raise the issue of ownership and responsibilities, and it often has been difficult to persuade land users to carry out voluntary maintenance after subsidized construction. Thus policies related to water harvesting can have important implications regarding investments and sustainable livelihoods in dryland agriculture.

These examples illustrate that effective water harvesting is much more than engineering design. It involves agronomic considerations, including fertility management; it offers possibilities for climate change adaptation through response-management; it has a bearing on land degradation and ecosystems func-tion; it is dependent on the creativity and commitment of land users, and closely connected to social security and livelihoods policy in the drylands. This leads us to propose the term 'water harvesting plus' (WH+) to signify the importance of including these crucial elements.

Concluding remarks

A considerable body of knowledge has been built up about water harvesting in Sub-Saharan Africa – and elsewhere – over the last quarter of a century. It can be argued that there are still gaps regarding technical performance, extent of systems and technology design, but this should not deter the further research, refinement and promotion of water harvesting as a tool to combat hunger in the drier parts of Africa. Water harvesting systems, when they perform well, turn runoff into an ally and capture sediment simultaneously. There is a solid case for considering water harvesting as having large unexploited potential in terms of production, livelihoods and the environment. Water harvesting is not foolproof; it ultimately depends on rainfall – and on human ingenuity to manipulate systems. But in the light of climate change and increasing risks in rainfed farming in Sub-Saharan Africa, effective water harvesting taking into account broader aspects beyond conventional engineering concerns is one of the best options for sustainable intensification of production.

Note

1 This chapter has been developed from a World Bank Water Paper entitled 'Improving Water Management in Rainfed Agriculture: Issues and Options in Water-Constrained Production Systems' by Scheierling et al. (2012); and also from a background paper prepared for that report by Critchley (2009).

References

Agarwal, A. and Narain, S. (eds) (1997) *Dying Wisdom: Rise, Fall and Potential of India's Traditional Water Harvesting Systems*, Centre for Science and Environment, New Delhi, India.

Agarwal, A., Narain, S. and Khurana, I. (eds) (2001) *Making Water Everybody's Business: practice and policy of water harvesting*, Centre for Science and Environment, New Delhi, India.

Barron, J. (2009) *Rainwater Harvesting: A Lifeline for Human Well-being*, Report for United Nations Environment Programme (UNEP) prepared by Stockholm Environment Institute (SEI).

Boers, T. M. and Ben-Asher, J. (1982) 'Review of rainwater harvesting', *Agricultural Water Management*, vol. 5, pp. 145–158.

Borst, L. and de Haas, S. A. (2006) 'Hydrology of Sand Storage Dams: A case study in the Kiindu catchment, Kitui District, Kenya', MSc thesis Hydrology, VU University, Amsterdam.

Bridges, M. E., Hannan, I. D., Oldeman, L. R., Penning de Vries, F. W. T., Scherr, S. J. and Sombatpanit, S. (2001) (eds) *Response to Land Degradation*, Oxford and IBH, New Delhi.

Briggs, D. J. (1986) 'Runoff farming in rural arid lands', an *Applied Geography* 'theme' volume, vol. 6, no. 1, pp. 5–81.

Chambers, R. (1983) *Rural Development: putting the poor first*, Longman, London.

Critchley, W. R. S. (1987) 'Some Lessons from Water Harvesting in Sub-Saharan Africa', Report from a workshop held in Baringo, Kenya, 13–17 October 1986, World Bank, Eastern and Southern Africa Projects Department, Washington D.C.

Critchley, W. R. S. (1991) *Looking After Our Land – Soil and Water Conservation in Dryland Africa*, Oxfam Publications, UK.

Critchley, W. (1999) 'Food-for-work and rainwater harvesting: Experience from Turkana District, Kenya in the 1980s', in: D. W. Sanders, P. C. Huszar, S. Sombatpanit and T. Enters (eds), *Incentives in Soil Conservation: from theory to practice*, World Association of Soil and Water Conservation, Oxford and IBH Publishing, New Delhi.

Critchley, W. R. S. (2009) 'Soil and water management techniques in rainfed agriculture', Background paper prepared for Water Anchor, The World Bank, Washington D.C.

Critchley, W. R. S. (2010) *More People, More Trees: environmental recovery in Africa*, Practical Action Publishing, Rugby, UK.

Critchley, W. R. S., Negi, G. and Brommer, M. (2008) 'Local Innovation in Green Water Management', in: Bossio D. and Geheb K., *Conserving Land, Protecting Water*, CABI Publishing, Oxford, pp. 107–119.

Critchley, W. R. S., Reij, C. and Seznec, A. (1992) 'Water harvesting for plant production, Volume II: Case studies and conclusions for Sub-Saharan Africa', World Bank Technical Paper Number 157, Africa Technical Department Series, Washington D.C.

Critchley, W. R. S. and Siegert, K. (1991) *Water Harvesting: A Manual for the Design and Construction of Water Harvesting Schemes for Plant Production*, FAO, Rome.

Dercon, S. (2002) 'Income risk, coping strategies, and safety nets', *The World Bank Research Observer*, vol. 17, no. 2, pp. 141–166.

Douglas, M. (1994) *Sustainable use of agricultural soils: A review of the prerequisites for success or failure*, Development and Environment Report, No. 11, University of Berne, Switzerland.

Falkenmark, M. (1995) 'Land-water linkages: a synopsis', in: Land and water integration and river basin management, *FAO Land and Water Bulleting*, no. 1, pp. 15–17. FAO, Rome.

Falkenmark, M., Fox, P., Persson, G. and Rockström, J. (2001) *Water Harvesting for Upgrading of Rainfed Agriculture: Problem Analysis and Research Needs*, SIWI Report 11, SIWI, Stockholm.

FAO (1994) 'Water harvesting for improved agricultural production', Proceedings of the FAO Expert Consultation, Cairo, Egypt, November 1993.

Fereres, E., and Soriano, M. A. (2007) 'Deficit irrigation for reducing agricultural water use', *Journal of Experimental Botany*, vol. 58, no. 2, pp. 147–159.

Finkel, M., Erukudi, C. and Barrow, E. (1987) *Turkana Water Harvesting Manual*, Mimeo, Ministry of Agriculture, Nairobi, Kenya.

Hilhorst, T. and Muchena, F. (eds) (2000) *Nutrients on the Move: Soil fertility dynamics in African farming systems*, International Institute for Environment and Development, London.

Howell, T. A. (1990) 'Grain dry matter yield relationships for winter wheat and grain sorghum—southern high plains', *Agronomy Journal*, vol. 82, pp. 914–918.

Hudson, N. W. (1971; 1981, 2nd edition) *Soil Conservation*, Batsford, London.

Hudson, N. W. (1987) 'Soil and water conservation in semi-arid areas', *Soils Bulletin*, no. 57, Food and Agriculture Organisation (FAO), Rome.

Hudson, N. W. (1991) 'A Study of the Reasons for the Success or Failure of Soil Conservation Projects', *Soils Bulletin*, no. 64, Food and Agriculture Organisation (FAO), Rome.

Hurni, H. ('with the assistance of an international group of contributors') (1996) *Precious Earth. From Soil and Water Conservation to Sustainable Land Management*, International Soil Conservation Organisation and Centre for Development and Environment, University of Berne.

IFAD (1992) *Soil and Water Conservation in Sub-Saharan Africa: Towards Sustainable Production by the Rural Poor*, Report prepared by the Centre for Development Cooperation Services, VU University, Amsterdam for International Fund for Agricultural Development, Rome.

IPCC (2007) *Fourth Assessment Report*, Intergovernmental Panel on Climate Change, Geneva, Switzerland.

Kerr, J. and Sanghi, N. K. (1992) 'Indigenous Soil and Water Conservation in India's Semi-Arid Tropics', *Gatekeeper Series*, no. 34, International Institute for Environment and Development, London.

Koohafkhan, P. and Stewart, B. A. (2008) *Water and Cereals in Drylands*. Earthscan, London.

Liniger, H. P., Mekdaschi Studer, R., Hauert, C. and Gurtner, M. (2011) 'Sustainable Land Management in Practice – Guidelines and Best Practices for Sub-Saharan Africa', TerrAfrica, World Overview of Conservation Approaches and Technologies (WOCAT) and Food and Agriculture Organisation of the United Nations (FAO).

MoALD (1984) *Interim Report Baringo Pilot Semi-Arid Area Project*, Ministry of Agriculture and Livestock Development, Government of Kenya.

Molden, D. (ed.) (2007) *Water for Food, Water for Life: A Comprehensive Assessment of Water Management in Agriculture*, Earthscan, London and International Water Management Institute, Colombo.

Narayana, V. V. D and Babu, R. (1985) *Soil and Water Conservation in Semi-arid Regions of India*, CSWCRTI, Dehradun, India.

NAS (1974) *More Water for Arid Lands: Promising Technologies and Research Opportunities*, National Academy of Sciences, Washington D.C.

Ngigi, S. N. (2003) *Rainwater Harvesting for Improved Food Security: Promising technologies in the greater horn of Africa*, Greater Horn of Africa Rainwater Partnership (GHARP) and Kenya Rainwater Association (KRA), Nairobi, Kenya.

Oweis, T. and Hachum, A. (2009) 'Water harvesting for improved rainfed agriculture in the dry environments', in S. P. Wani, J. Rockström and T. Oweis (eds), *Rainfed Agriculture: unlocking the potential*, CABI, Oxford.

Oweis, T., Hachum, A. and Bruggeman, A. (eds) (2004) *Indigenous Water Harvesting Systems in West Asia and North Africa*, ICARDA, Aleppo, Syria.

Oweis, T., Hachum, A. and Kijne, J. (1999) 'Water Harvesting and Supplementary Irrigation for Improved water Use Efficiency in Dry Areas', *SWIM Paper no. 7*. ICARDA/ IWMI.

Oweis, T. Y., Prinz, D. and Hachum, A. Y. (2012) *Rainwater Harvesting for Agriculture in the Dry Areas*, ICARDA, CRC Press/Balkema, Leiden, the Netherlands.

Pacey, A. and Cullis, A. (1986) *Rainwater Harvesting: the collection of rainfall and runoff in rural areas*, IT Publications, London.

Perry, C., Steduto P., Allen, R. G. and Burt, C. M. (2009) 'Increasing productivity in irrigated agriculture: Agronomic constraints and hydrological realities', *Agricultural Water Management*, vol. 96, no. 11, pp. 1517–1524.

Rochette, R. M. (1989) *Le Sahel en Lutte Contre la Desertification: Leçons d' éxperiences*, GTZ, Eschborn.

Rockström, J. (2000) 'Water resources management in smallholder farms in eastern and southern Africa: An overview', *Physics and Chemistry of the Earth*, vol. 25, no. 3, pp. 275–283.

Rockström, J. and Falkenmark, M. (2000) 'Semiarid crop production from hydrological perspective: Gap between potential and actual yields', *Critical Reviews in Plant Sciences*, vol. 19, no. 4, pp. 319–346.

Scheierling, S. M., Critchley, W. R. S., Wunder, S. and Hansen, J. W. (2012) 'Improving water management in rainfed agriculture: Issues and options in water-constrained production systems', Water Paper, Water Anchor. The World Bank, Washington D.C.

Shaxson, T. F., Hudson, N. W., Sanders, D. W., Roose, E. and Moldenhauer, W. C. (1989) *Land Husbandry. A Framework for Soil and Water Conservation*, Soil and Water Conservation Society, Ankeny, Iowa.

van Steenbergen, F., Lawrence, P., Haile, A. M., Salman, M. and Faurès, J. M. (2010) 'Guidelines on spate irrigation', FAO Irrigation and Drainage Paper 65.

Toulmin, C. (2009) *Climate Change in Africa*, in association with International African Institute, Royal African Society and Social Science Research Council, Zed Books, London and New York.

Vlek, P. L. G., Bao Le, Q. and Tamene, L. (2008) 'Land Decline in Land-Rich Africa: A Creeping Disaster in the Making', Consultative Group on International Agricultural Research, Centre for Development Research (ZEF), University of Bonn, Germany.

Wani, S. P., Rockström, J. and Oweis T. (eds) (2009) *Rainfed Agriculture: Unlocking the potential*, CABI, Oxford.

WCED (1987) *Our Common Future*, World Commission on Environment and Development. Oxford University Press, Oxford.

Whiteman, P. (1981) *Sorghum and Millet Agronomy Investigations in Eastern Province*, Technical Report, Katumani. Scientific Research Division, Ministry of Agriculture, Republic of Kenya.

WOCAT (2007) *Where the land is greener: case studies and analysis of soil and water conservation initiatives worldwide*. Liniger, H. and Critchley, W (eds), CTA, FAO, UNEP, CDE, Bern.

World Bank (2011) *Climate-Smart Agriculture: A Call to Action*, World Bank, Washington D.C.

Zimmerman, F. J. and Carter, M. R. (2003) 'Asset Smoothing, Consumption Smoothing and the Reproduction of Inequality under Risk and Subsistence Constraints', *Journal of Development Economics*, vol. 71, pp. 233–260.

Chapter 3

A review of the recent literature on water harvesting in Sub-Saharan Africa

Jetske Bouma, William Critchley and Jennie Barron

Introduction

This chapter constitutes a review of the recent peer-reviewed literature on water harvesting in Sub-Saharan Africa. It attempts to analyse what researchers and academics have studied and published about water harvesting over the last 20 years in the light of renewed interest in the topic. When water harvesting was first proposed as a possible answer to the droughts of the 1970s and 1980s, there were a small number of key publications which, in the pre-digital age, were to be found as well-thumbed copies on enthusiasts' shelves. *More Water for Arid Lands* (NAS, 1974) was one of the first. This slim volume drew attention to the potential of 'runoff agriculture': a new concept to many in the field of agricultural development world. To others familiar with traditions in semi-arid areas it simply conferred a name on, and legitimacy to, an age-old practice. Boers and Ben-Asher (1982) produced perhaps the first general literature review of the topic. They noted:

> During the past 25 years water harvesting has been receiving renewed attention. We found roughly 170 articles ... which had appeared between 1970 and 1980: about three quarters of them originated from the USA.

Most of the other papers they reviewed were from India, Israel and Australia. Africa was notably absent from their article – perhaps ignored, but certainly poorly represented in the literature. It was not until Pacey and Cullis (1986) produced their seminal book on rainwater harvesting that Sub-Saharan Africa's experience and potential featured prominently. While many of the 303 references cited (a mixed bunch of journal articles, project reports and memos) were not location-specific, only 46 were directly Africa-related. Critchley et al. (1994) reviewing indigenous soil and water conservation globally, albeit with 'a deliberate focus on moisture deficit regions of sub-Saharan Africa', cite around 60 references which are directly related to Africa (just less than half those reviewed) indicating the growing body of writing about water harvesting and related topics on the continent. But where has the published research taken us since?

Figure 3.1 shows the number of papers published per five-year period for specific 'Web of Science' (ISI Web of Knowledge v.5.6) search terms.

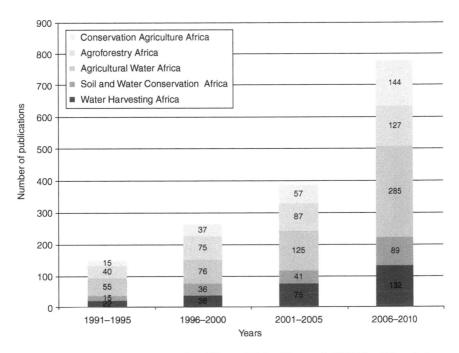

Figure 3.1 Number of publications for different 'Web of Science' (ISI Web of Knowledge v.5.6) search terms Sub-Saharan Africa

However, the different categories are not mutually exclusive, thus the publications under 'Agricultural Water Africa' are likely to include (at least) some of the publications under 'Water Harvesting Africa'; and 'Soil and Water Conservation Africa' and 'Water Harvesting Africa' are also likely to overlap in some instances.

For all categories, the number of publications is increasing, and the largest absolute increase from 1991 to 1995, and 2006 to 2007, is 'Agricultural Water Africa'. But the greatest *relative* increase between these date categories has been 'Conservation Agriculture' followed by 'Water Harvesting'. This is especially significant as these are two disciplines that have attracted special attention in recent years with soil health, rainfed agriculture and a changing climate becoming increasing concerns.

Regarding the disciplinary background of the various papers, Figure 3.2 indicates that most papers have an agronomic, water resources or ecology component; that many have a soil science, forestry, meteorological studies or plant sciences background; and that some mention engineering or economical aspects of the technologies concerned. Clearly, agroforestry studies address forestry-related topics, and research in the field of conservation agriculture is characterized by a more ecological approach. Studies on agricultural water and water harvesting focus more on water resources and meteorological issues, while research on soil and water conservation generally take a more soil-sciences-focused ecological approach.

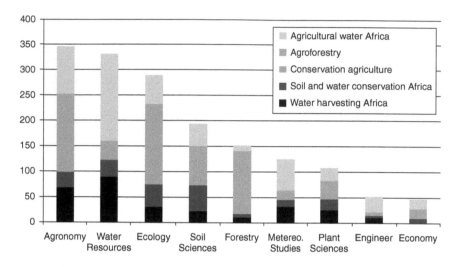

Figure 3.2 Categorization by disciplinary content: literature search through 'Web of Science' (ISI Web of Knowledge v.5.6)

In line with the theme of this book we concentrate our analysis on the water harvesting literature, with some attention given to soil and water conservation articles where relevant. The reason for only considering the published literature is that this literature is more easily accessible and since it has been peer-reviewed its quality has been checked. We were especially interested in papers presenting the impacts of rainwater harvesting and those that discuss the wider context in which rainwater harvesting investments take place. Thus, this chapter prepares the ground for the country-specific chapters, which go more deeply into technical aspects and environmental characteristics. Also, we concentrated on English-language papers: we are aware that a supplementary French-language could have added value to this review, as more Sahelian references would have been found, but it is unlikely to have made a significant difference to our analysis as many related references are included in the English-language literature as well. The main aim of the chapter is to achieve an overview of the published literature on water harvesting in Sub-Saharan Africa over the last 20 years, in order to attain a good understanding of the main findings and to identify some of the main gaps.

General findings

When reviewing the literature it immediately becomes apparent that a considerable amount of research has been carried out over the last 20 years. Since 1990, more than 300 studies have been published – and countless more when considering conference proceedings, project documents and so forth. Most of the studies herald the potential benefits of water harvesting, while discussing the importance of increasing 'green water productivity' for better food security and more resilient livelihoods (e.g. Hoff et al., 2010; Rockström et al., 2010). Many studies also illustrate these benefits: crop yields double (e.g. Fox and Rockström, 2003; Roose et al., 1999) or even triple (e.g. Barry et al., 2008; Fox et al., 2005; Sawadago, 2011), degraded lands become more productive or are simply brought back into production (Kabore and Reij, 2004) and domestic water availability is improved (Kahinda et al., 2007). However, few studies translate these benefits into improved food security or better livelihoods; nevertheless, those that do generally indicate that rural livelihoods improve because of water harvesting (Bekele and Drake, 2003; Hatibu et al., 2006; Kusangaya, 2006; Moges et al., 2011; Mutekwa and Hanjra et al., 2009; Shiferaw and Holden, 2001). In most cases the farm households that benefited from water harvesting did not invest in the systems directly themselves: except for few traditional or pioneering cases, most of the literature concerns farmers being supported in their water harvesting investments, although farm-households often contribute their labour, sometimes at considerable opportunity costs. Possibly because of the subsidized nature of many water harvesting investments, relatively few studies compare benefits with costs, or calculate investment returns. Generally, it is assumed that farmers are unable to invest in water harvesting, and that subsidies are required (and justified) to achieve adoption and uptake.

The multi-disciplinary nature of most of the research on water harvesting is another aspect that captures attention. The positive side of this is that many studies address both contextual, technical and farming-practice-related factors; but the lack of a common, integrated framework also complicates the analysis of impacts and enabling conditions across sites and techniques. Social science perspectives are under-represented, with most studies adopting an agronomic, hydrological or soil sciences viewpoint. The scale of analysis is mostly field/farm level, although a few studies consider impacts in the broader watershed as well; for example, Wisser et al. (2010) at global level and Lasage et al. (2008). Given the relatively limited impact of small and dispersed water harvesting investments on streamflows this is not surprising, but an increasing number of studies show that when these small water harvesting interventions become numerous they can have an aggregate impact downstream (e.g. Bouma et al., 2011; Pachpute et al., 2009). The plot level focus of most water harvesting activities in Sub-Saharan Africa implies that those involved usually comprise individual farmers, rather than village communities or farmer groups. Exceptions are found in Mazzucato et al. (2001) and Baipheti et al. (2009) who do assess water harvesting at community level, and indicate that group approaches can generate higher overall impacts when implemented well.

Water harvesting technologies and their impacts

The water harvesting technologies most commonly described in the literature are *zaï* (planting pits in Burkina Faso's *Moré* language; *tassa* in Niger's *Hausa*), stone lines, *demi-lunes*, household ponds and small reservoirs, permeable rock dams and spate irrigation/floodwater harvesting, and combinations with agronomic measures including contour ploughing and manuring. Most techniques are discussed as part of a mixture; that is, *zaï* and stone lines or combinations of bunds, stone lines and contour ploughing. However spate irrigation/floodwater harvesting, small dams, household and community ponds are usually discussed independently.

Considering the technical design of water harvesting, most of the technologies referred to in the literature are standard with little attention given to innovations or modifications. The innovations that *are* cited include farmer initiatives; for example, road runoff harvesting (Critchley and Mutunga, 2003) and 'gully gardens' (Critchley et al., 2008), combinations of different water harvesting techniques (Hatibu et al., 2003; Pachpute et al., 2009), and the use of pumps for supplementary irrigation in the context of small dams and household ponds (Moges et al., 2011).

The most commonly reported impacts of water harvesting investments are crop yield improvements. Typically, studies report higher returns on experimental plots compared to those on farmers' fields; but many experiments are conducted on-farm and it is difficult to clearly distinguish between researcher-controlled on-farm trials, farmers' experiments and their normal practices. Generally, studies report crop yield improvements within the range 2–200 per cent but the range differs substantially per technique and per crop. Crops are usually the common subsistence cereals of sorghum, millet and maize, but when water harvesting allows for supplementary irrigation, farmers often start growing vegetables as well. This is reported to substantially improve household consumption (Tesfaye et al., 2008) and farm income (Hatibu et al., 2006), but even when farmers don't change their cropping patterns, the highest ranges of yield improvements are still reported when water harvesting results in supplementary irrigation, for example through household ponds: different studies assessing the impacts of household ponds report yield improvements in the range 44–170 per cent (Fox and Rockström, 2000, 2003; Barron and Okwach, 2005; Fox et al., 2005; and Hatibu et al., 2006).

The crop yield improvements reported for planting pits (*zaï*) are much lower: Maatman et al. (1998) report yield increases of 2–19 per cent and Roose et al. (1999) cite increases of 19–48 per cent. A few studies assessing the impacts of planting pits compare good yields with planting pits, to a situation where no crop production is possible without water harvesting (e.g. Roose et al., 1999; Hassane et al., 2000; Kabore and Reij, 2004). This is not an unjustified comparison since in many Sahelian countries crop production on highly degraded soils (e.g. the *zipele* of Burkina Faso and the *fako* of Niger) without deep and wide planting pits is simply not possible. The importance of soil fertility and manure application is underlined in many studies: crop yields increase up to 10 times when manure is added as well (Fox and Rockström, 2003; Fox et al., 2005; Hassane et al., 2000; Roose et al., 1999). As Stroosnijder (2009) argues, most farmers (and policy makers) are only aware of the drought component of low crop yields, forgetting the crucial role played by lack of nutrients in degraded soils. However there is clearly a difference between water harvesting based on collection of localized runoff and the 'self-fertilizing' systems of floodwater harvesting which replenish the cultivated land with deposited sediments – the same basic process that develops alluvial soils (Steenbergen et al., 2011).

The impacts of stone lines are comparable to those of planting pits, Alemayu et al. (2006) reporting improvements of 6–17 per cent and Zougmorë et al. (2004) of 12–18 per cent. As in the case of planting pits, the incremental yield resulting from water harvesting is much higher when investments are simultaneously made in soil fertility. Barry et al. (2008) report 13–100 per cent improvements for different water harvesting measures (grass strips, stone lines, *zaï*) and Sawadogo et al. (2011) cite 60–240 per cent improvements for planting pits, stone lines and additional manure. Looking more broadly at *in situ* conservation measures as well as water harvesting, combinations of technologies usually lead to higher yield improvements: mixtures of earth bunds and grass strips, contour ploughing and infiltration pits, are good examples. Motsi et al. (2004) report increments of 53–127 per cent for combinations of tied ridges and *fanya juu* terraces (contour earth bunds with infiltration trenches on the lower side).

Although the importance of water harvesting for improved food security is much heralded (e.g. Hanjra et al., 2009), studies actually demonstrating this at household level resulting from water harvesting investment are few. Tesfaye et al. (2008) indicate that access to small-scale irrigation, not surprisingly, significantly enhances food security and Fox et al. (2005) find, in a comparative study of Burkina Faso and Kenya, that water harvesting for supplemental irrigation improves self-sufficiency in staple food production as well. However, crop yield improvements *per se* do not necessarily lead to improved food security, as the latter also depends on timing, for example whether food security improves in times of drought (see Ericksen et al., 2011). Kato et al. (2011) address the role of water harvesting in reducing climatic risks, indicating that water harvesting investments help reduce agricultural production risks, but that impacts differ significantly across regions and technologies. Barbier et al. (2009) show that drought risks have induced farmers to invest in planting pits and stone lines in Burkina Faso, and Deressa et al. (2009) demonstrate something similar for Ethiopia – but assessments of the benefits of water harvesting in years of extreme drought are scarce. Amongst others, Roose et al. (1999), Hassane et al. (2000) and Barron and Okwach (2005) emphasize that water harvesting improves crop yields in years of below average rainfall, but there are no studies that demonstrate conclusively the extent of the contribution water harvesting could make in the face of predicted climate change in Sub-Saharan Africa.

Environmental effects

Water harvesting can also have important environmental effects: most systems contribute to environmental sustainability by reducing soil erosion (e.g. Hengsdijk et al., 2005; Roose et al., 1999) and many also contribute in terms of increased biomass, biodiversity and soil fertility (e.g. Reij and Smaling, 2008; Stroosnijder, 2009; Sawadogo, 2011). Redistributing rainfall in time and space, as achieved by water harvesting, may have potential downstream or in-stream impacts on water resources and water-dependent ecosystem services and livelihoods. These impacts may be desired or undesired. Clearly, bridging dry spells to enhance crop production, and improve land, water and labour productivity, is positive. However, it may affect downstream users as water flows can be influenced due to large-scale upstream water harvesting adoption. As a rule of thumb, hydrological impacts can be substantial in shorter time-spatial scales, whereas they tend to decrease with increasing time step and/or spatial scale of assessment (e.g. Schulze, 2000). According to van Noordwijk et al. (2004) effects of changing land use in terms of water flows may not be felt beyond approximately 1000 km from the site of impact. Andersson (2011) estimated for southern Africa river basins (1.8×10^6 km^2) both *in situ* moisture conservation and water harvesting had insignificant impact on downstream flows, even when water harvesting was combined with high standards of nutrient management for plant growth.

Impact is clearly affected by the area of the basin/watershed which is under water harvesting. Wisser and colleagues (2010) estimated that the implementation of small dams could on average reduce downstream flow by 2 per cent. Thus, implementing water harvesting within scattered farmers' fields can be expected to have even less impact on mean annual flows. There are yet only a few robust assessments of water harvesting interventions that provide data on downstream, and within stream, water partitioning and flows at the meso-scale (10–10,000 km^2), the scale at which most integrated resource land and water management is put into practice. In a meso-scale watershed in semi-arid Burkina Faso with 700 mm/ year long-term average rainfall, the introduction of microcatchment water harvesting was modelled without significant impacts on downstream flows when 70 per cent of the 1000 km^2 watershed was directly affected by this management practice (after Barron, 2012). Water harvesting interventions can potentially have substantial impacts in reducing sediment flows, as the water partitioning changes and more rainfall infiltrates at plot/field scale (Andersson et al., 2009). For example Magombeyi and Taigbenu (2011) have shown that soil and water conservation/ water harvesting could reduce water flows from a 534 km^2 semi-arid catchment in the Olifants basin, South Africa, by 14.3–19.8 per cent, while sediment loss was reduced by 41–46 per cent.

Apart from the impacts on biomass, water partitioning, flows and sediments, there are few published research findings or modelled predictions of other possible environmentally positive or negative impacts of large scale water harvesting adoption. Vohland and Boubacar (2009) reviewed various aspects and the evidence of impact relating to micro and macro fauna, and flora, at various scales – from plot to landscape – was fairly small. As their synthesis concludes, there is little systematic understanding of the impacts of water harvesting in

Sub-Saharan Africa, and in view of the current focus on broader sets of effects on ecosystem services resulting from agricultural practices, this is an important and urgent area for further research.

Returns to water harvesting investment and implementation

As the preceding sections have indicated, the positive impacts of water harvesting can be substantial, with crop yield improvements of up to 200 per cent, environmental benefits and in some cases climate change risk reduction effects as well. Often, these benefits are not valued economically, but the few studies that do assess economic benefits indicate that farm household income or consumption improves: for example, Magombeyi and Taigbenu (2011) demonstrate income benefits for smallholder farmers in South Africa from between US$4 and US$270 under conventional rainfed to between US$233 and US$1140 under supplementary irrigation depending on household typology; and for Ethiopia several studies indicate significant benefits of water harvesting for household income as well (see for example Shiferaw and Holden, 2001; Bekele and Drake, 2003; and Amsalu and de Graaf, 2007). Hatibu et al. (2006) evaluate the costs and benefits of water harvesting investments for East Africa, indicating that

> rainwater harvesting for production of paddy rice paid most with returns to labour of more than US$ 12 per person-day invested, benefits being very high due to the fact that without rainwater harvesting it is not possible to produce paddy in the study area and rainfed sorghum realizes a return to labour of only US$ 3.7 per person-day.

Most economic studies analysing water harvesting investments do not, however, really focus on evaluating benefits, but instead try to explain why, with relatively high benefits, water harvesting adoption rates remain relatively low. An important reason constitutes implementation and maintenance costs: Shiferaw and Holden (2001) point out that water harvesting can incur substantial opportunity costs in terms of land use. Water harvesting technologies may take some land out of production (e.g. where bunds are built within fields) and if farmers only have small plots this can be a substantial cost. An implicit assumption with regard to labour costs is that the farm-household's opportunity costs of labour are zero, especially in the dry season when agricultural activity is low. In many Sub-Saharan Africa regions the opportunity costs of labour are increasing; however, Fox et al. (2005) showing that in regions with higher opportunity costs of labour the returns to investment are (obviously) less. Also, Moges et al. (2011) review the returns to investment of household water harvesting structures in Ethiopia and conclude that due to high opportunity costs of labour, and disappointingly low benefits (caused by high seepage and evaporation losses in household ponds), returns to investments are in many cases insufficient to cover investment costs.

This actually relates to a finding from the broader literature on water harvesting investments which tries to explain why smallholder farmers may be reluctant to invest in water harvesting: Walker and Ryan (1990) show, for example, that in India, low farm gate prices, uncertain revenues and increasing opportunity costs of labour make investments in rainfed agriculture relatively unattractive, especially when compared to investments in supplementary irrigation that have a higher benefit–cost ratio. Clearly, this benefit–cost ratio may vary over time: traditionally, farmers might have invested in water harvesting measures (see Chapter 9 for two examples from Sudan), but over time the opportunity costs of labour might change. Subsidization improves the benefit–cost ratio to the farmer, but farmers still need to maintain investments in the long run. Hatibu et al. (2006) and Moges et al. (2011) argue that this is more likely to happen when water harvesting makes it possible for farmers to shift production to higher value, more water-demanding cash crops, although research in more arid regions suggests that subsistence farmers in mixed crop-livestock systems maintain water harvesting investments as well (Reij and Smaling 2008). Amsalu and de Graaff (2007) indicate that, in Ethiopia, the perceived profitability of water harvesting is an important determinant of future maintenance; higher returns to investment affect maintenance positively, and off-farm employment has a negative effect.

A further key reason why farm households may not maintain water harvesting investments is because they expect the external investor to do this for them. Moges et al. (2011) indicate that in Ethiopia farmers expect the government to maintain investments and in India, Kerr et al. (2002) show that maintenance is less in projects where subsidization is high. Maintenance might also be less when farmers are not consulted during water harvesting implementation: Liu et al. (2008) explain how participatory water harvesting – that is when water harvesting implementation is planned, designed and implemented together with the farm households and/or wider community – this does not only increase local commitment but also helps overcome informational problems by making use of stakeholder knowledge regarding local context and the specific problems that need to be addressed. Thus, investments are better located, resulting in lower costs and higher returns. Sturdy et al. (2008) illustrate that participatory approaches and farmer-driven experimentation are essential for ensuring water harvesting relevance and adaptation to the local context, and Critchley (2009) discusses several studies that suggest that top-down technocratic approaches don't work and that farmers need to be involved in technology development and implementation.

The practice of many water harvesting projects in Africa seems rather non-participatory, however: Moges et al. (2011) describe how Ethiopia's water harvesting programme is top-down, technocratic and non-participatory. Bewket (2007) indicates that even in a project that presented itself as participatory, farmers were not truly consulted but merely persuaded to accept water harvesting investments on their land. The limited attention to participatory and community-based approaches in parts of Africa – especially Ethiopia – is surprising given the experiences in India that indicate that participatory approaches are much more successful (Kerr et al., 2002) and efficient in terms of water harvesting investment costs. Nevertheless one of the most celebrated successes of water harvesting in Sub-Saharan Africa, namely the combination of stone bunds and *zaï* planting pits in Burkina Faso, was founded in participatory processes, and is renowned for its pioneering people-centred approach (Critchley, 2010). Barron and Noel (2011) explicitly compare water harvesting approaches between Africa and India confirming that the 'soft component' of stakeholder participation and farmers' knowledge is indeed crucial for water harvesting success. Baiphethi et al. (2009) discuss the role of local institutions for successful water harvesting implementation, indicating that in South Africa farmers' groups played an important role in coordinating labour efforts and making water harvesting a success. In the following we further discuss the factors determining water harvesting uptake and upscaling, elaborating the role of contextual, institutional and capacity-related factors in water harvesting adoption and spread.

Water harvesting uptake and upscaling – enabling factors and constraints

Drechsel et al. (2010) conclude from an expert meeting on water harvesting uptake in Africa that no comprehensive assessments of the factors determining effective uptake exist: although the biophysical requirements for water harvesting are well described, information about the socio-economic, cultural and institutional conditions for uptake and upscaling is lacking. A seminal study by Feder et al. (1985) on smallholder adoption of agricultural technologies suggests that price support schemes, food taxes and aid programmes influence adoption to a large extent. Suri (2011) convincingly shows that smallholder farmers in Sub-Saharan Africa are constrained in their adoption of new technologies because of 'missing markets': a substantial share of farmers have potentially high returns to investment but also high adoption costs because they lack good market access. Duflo et al. (2011) illustrate something similar for the case of fertilizer, which also has low adoption rates in Sub-Saharan Africa. Here, lack of access to financial services (credit and savings) constrain farmers in buying fertilizer when they need it, the analysis indicating that selling fertilizer coupons to farmers immediately post-harvest is much more effective than subsidizing fertilizer throughout the year. Also in the case of water harvesting adoption, missing or malfunctioning markets seem to seriously undermine uptake: Tabo et al. (2011) discusses how combined interventions in micro-dosing of fertilizer in planting pits and establishment of crop storage and savings systems (e.g. the *warrantage* system) helped to substantially improve farmer incomes in the Sahel.

The problem of missing or malfunctioning markets is especially relevant to the poor: Misselhorn (2005) concludes that restricted market access is one of the main factors explaining food insecurity in Southern Africa, and in general poor household are usually constrained in their access to markets, which is an important reason why they remain poor (Duflo et al., 2011). These households are also faced by other constraints, including lack of assets, entitlements and property rights, but it is outside the scope of this chapter to fully discuss the role of poverty in influencing water harvesting adoption and uptake. Still, Barbier (1990) illustrates the crucial role of land tenure security for adoption of conservation measures, and Shiferaw and Holden (2001) and Kabubo-Mariara (2007) analyse the role of land tenure in water harvesting uptake.

Finally, lack of information and awareness is another reason explaining low adoption rates: farmers are often not aware of different water harvesting technologies and do not have the education or access to informational networks to acquire this knowledge themselves. Critchley (2009) and Critchley et al. (1992) discuss the importance of agricultural extension and the role of farmer knowledge and farmer-to-farmer extension in water harvesting adoption, underlining the important role of capacity building in uptake. The World Overview of Conservation Approaches and Technologies (WOCAT: www.wocat.net) has been monitoring the implementation of soil and water conservation (including water harvesting) technologies and has built a large database of experiences over time. Critchley (2009) reviews this database amongst other literature, concluding that

> [a]mongst the points highlighted [for limited uptake] are: an over-emphasis on engineering approaches; failure to train farmers in simple surveying techniques; the use of catchments/watersheds as the unit of intervention rather than the community; ignoring indigenous knowledge and traditions; and failure to develop partnerships and alliances. Furthermore there is the need to look upon growth in population density as a potential opportunity; the importance of considering livestock while planning improvements in small-scale farming; the danger of becoming obsessed with 'quick-fix' technical solutions; an obsession with the benefits of trees; the misuse of incentives to participating communities, especially the uncritical use of food-for-work, inducing a culture of dependence; and the unrealistic over-funding of small projects which cannot realistically be replicated on a large scale.

Finally, water harvesting must be a 'felt need'. This requires farmers to either acknowledge that their yields are low and could be better, or to understand the problem of climate change and/or land degradation, and realize that there is something they can do to ameliorate their potential losses. Barbier et al. (2009) find that in northern Burkina Faso farmers report that most new techniques have been adopted because of growing land scarcity and newly emerging markets, rather than the threat of climate change. Kato et al. (2011) report that in Ethiopia half of the farmers surveyed had made some investments to adapt to climate change. Hassan and Nhemachena (2008) show that for the whole of Africa, 50 per cent of the surveyed farmers believe that temperatures are increasing and precipitation is becoming less, but only 20 per cent of the respondents believe that soil and water conservation could help them adapt. Reij and Smaling (2008) find that innovations are mostly crisis-driven, but that market opportunities play a crucial role. This latter point is also concluded by Hassan and Nhemachena (2008); again underlining that for adoption it is crucial that missing markets are addressed.

Discussion

Our review of the recent literature on water harvesting in Sub-Saharan Africa has uncovered some interesting and important findings. Some are hardly 'new' – for example the enduring lack of quantitative estimates of adoption in terms of number of farmers and area under water harvesting. Others reflect a better understanding of how water harvesting systems can be improved – an example is its association with soil fertility management or the involvement of farmers in water harvesting design and implementation to increase commitment and reduce informational costs. Gaps in the literature, and by inference current research, are also apparent. Little attention seems to have been paid to gender-related issues, for example. In this concluding discussion we identify the main issues that emerge from this review.

'Water harvesting plus'

A well-documented feature of water harvesting is that when combined with investments in soil fertility it generates the highest returns. This may be clear to researchers but this realization does not seem to have reached the water harvesting and rural development policy circles yet, thus the importance of 'WH+' as proposed in Chapter 2. Furthermore, different bodies of literature address different issues: soil fertility and degradation issues on the one hand, and drought and water issues on the other. In reality both are related and we would like to again stress the need for an integrated approach.

Assessing impacts

Related to the need to better integrate soil fertility issues in water harvesting implementation, the literature indicates that an integrated framework for assessing water harvesting impacts is lacking. Different disciplinary approaches result in different choices of parameters, making it difficult to compare impacts across sites. Another important observation relating to the disciplinary background of most water harvesting studies is that there are few socio-economic studies. This is an important lacuna, given the importance of contextual factors and household characteristics in explaining water harvesting uptake. Finally, there is need for multi-scale assessment of water harvesting, both in terms of spatial extent and in terms of types of technologies. Most assessment now focus on single technologies and/or plot level, whereas a systematic synthesis may help draw out specific and generic features, benefits and impacts of water harvesting, learning from multiple contexts and conditions.

Downstream impacts and environmental effects

The review indicates that there is a large knowledge gap on systematic hydrological impacts at various scales up to landscape and basin as well as other impacts on the local environment, including carbon storage and biodiversity. Given the potential aggregate impacts of water harvesting investments on downstream regions, it is important to account for the wider hydrological impacts and assess water harvesting impacts at basin scale. Not accounting for environmental benefits, like carbon storage, results in undervaluation of water harvesting benefits, with possibly underinvestment as a result.

Scale of intervention

Related to the point that most studies ignore basin level impacts, the scale of intervention is in most cases limited to the plot level as well. This is surprising, as there are several studies indicating that approaches that focus at community or watershed level are often able to generate higher returns. Targeting communities instead of individual farmers may help increase water harvesting benefits at the wider social and spatial scale, but also requires a better understanding of the local socio-economic and institutional context which in many of the current water harvesting interventions seems to lack.

Top-down approaches and subsidies

Despite the preponderance of participatory rhetoric in development circles, as we have noted the reality is that implementation of most water harvesting programmes in Africa still seems to be characterized by top-down, subsidized, household-targeted approaches with little room for true participation or knowledge exchange. Several studies point to the importance of consulting farmers to incorporate their knowledge and better target investments to farmer's needs. Clearly, this influences farmers' perceptions of costs and benefits and will increase the likelihood that water harvesting investments are made, and maintained. The big question seems to be how farmers can be supported in making water harvesting investments without distorting the incentive to maintain these investments in the long run. Here it is important to be realistic, and work from heterogenous cost and benefit perceptions, not the technical costs and benefits that don't take local constraints and conditions into account.

Returns to water harvesting investment

With regard to the costs and benefits of water harvesting investments, this review indicates that benefits are usually not economically valued, and that they are often not compared to water harvesting costs. Several studies indicate that the costs of water harvesting implementation and maintenance can be substantial, but are often ignored because project implementers assume zero opportunity costs. Given that the opportunity costs of labour in Sub-Saharan Africa seem to be increasing, water harvesting innovations that reduce manual labour requirements in cost efficient ways seem to be a promising approach. Another factor that affects costs and benefits perceptions is the certainty with which benefits are gained: if water harvesting benefits are substantial in above-average rainfall years but disappointing in below-average rainfall years, risk-averse farmers are less likely to adopt, especially if alternative climate risk reduction strategies like off-farm employment exist. Surprisingly, the review indicated that few studies explicitly assess the climate risk reduction effects and potential livelihood impacts of water harvesting investments, which is a serious knowledge gap indeed. Finally, the review showed that even when the returns to water harvesting are substantial, farmers may not adopt because they lack access to markets, extension and credit services, which greatly increases their adoption costs. Addressing the issue of these so-called 'missing markets' seems crucial for the uptake and upscaling of water harvesting in Sub-Saharan Africa, as several studies indicate.

The business case for water harvesting

Overall, there seems to be a business case for water harvesting, especially where ponding of water allows supplementary irrigation and when this allows farmers to (at least partly) shift their farming practices from the production of subsistence, to cash crops. Clearly, this is not always possible as it depends on the amount of water that can be harvested and whether farmers have good market access. The business case for water harvesting in subsistence farming in rainfed agriculture is more uncertain, studies being limited to crop yield improvement assessments without attention to investment costs or costing benefits of increase landscape protection and reduced sediment flows. While water harvesting is a *sine qua non* for subsistence in many areas of Africa, there still need to be questions asked about the best investment options. Also, with the likelihood of increasing droughts in the future, the question arises whether current water harvesting systems in dryland agriculture provide adequate benefits in poor rainfall years: the literature provides little guidance on this crucial issue.

Some issues for further research

Finally, some observations about what appears to be missing or underplayed in the literature – and what we believe is worthy of future research. While our review focused on impact and uptake, there are several apparent lacunae, and some have been mentioned already – for example the paucity of social science studies and the lack of an integrated framework for such analyses. Perhaps the most surprising omission is that there is little in the literature on climate change, and the specific role of water harvesting as an adaptation technology. Another area for potential research is the impact of early warning systems that could reduce risk through sharing climatic data: many water harvesting systems are flexible (in terms of catchment area and 'response cropping' strategies) and could be modified by farmers, if suitably forewarned. The role of machinery needs to be better investigated: is it possible to reduce labour costs while introducing mechanization that can be maintained and sustained? Land tenure, and more generally, rights to resources appears to have been under-examined in terms of water harvesting. Security of tenure is a prerequisite for investment in water harvesting, and what about increasing competition for catchment areas (e.g. roads from which runoff is harvested)? The role of fertility management has been high-lighted; also important is to identify the benefits to be derived in terms of improved agronomy and the resultant extra capacity of water harvesting to bridge dry spells. Which crops are best suited to the extremes of water harvesting – inundation and intermittent drought? What work is being carried out on this? Finally, there is no evidence of diachronic studies – taking a particular situation and following it through to monitor its evolution over time. These are all opportunities, and can help fill some current gaps to create the fabric of a convincing portfolio of future water harvesting research.

References

Alemayhu, M., Yohanessi, F. and Dubale, P. (2006) 'Effect of indigenous stone bunding (Kab) on crop yield at Mesobit-Gedeba, North Shoa, Ethiopie', *Land Degradation and Development*, vol. 17, pp. 45–54.

Amsalu, A. and de Graaff, J. (2007) 'Determinants of adoption and continued use of stone terraces for soil and water conservation in an Ethiopian highland watershed', *Ecological Economics*, vol. 61, pp. 294–302.

Andersson, J. C. M. (2011). *The potential impacts of enhanced soil moisture and soil fertility on smallholder crop yields in Southern Africa*. Doctoral dissertation. Federal Institute of Technology Zurich, Zurich, Switzerland.

Andersson, J. C. M., Zehnder, A. J. B., Jewitt, G. P. W. and Yang, H. (2009) 'Water Availability, Demand and Reliability of in situ Water Harvesting in Smallholder Rain-fed Agriculture in the Thukela River Basin, South Africa', *Hydrology and Earth System Sciences*, vol. 13, pp. 2329–2347.

Baiphethi, M., Vilhoen, M., Kundhlande, G., Botha, J. and Anderson, J. (2009) 'Reducing poverty and food security by applying in-field rainwater harvesting: how rural institutions made a difference', *African Journal of Agricultural Research*, vol. 4, no. 12, pp. 1358–1363.

Barbier, E. (1990) 'The farm-level economics of soil conservation – the uplands of Java', *Land Economics*, vol. 66, pp. 199–211.

Barbier, B., Yacouba, H., Karambiri, H., Zorome, M. and Some, B. (2009) 'Human vulnerability to climate variability in the Sahel: Farmers' adaptation strategies in northern Burkina Faso', *Environmental Management*, vol. 43, pp. 790–803.

Barron, J. (2012). *Soils as a water resource: some thoughts on managing soils for productive landscapes meeting development challenges*. (Wageningen UR Open Journal. Thematic Issue) http://library.wur.nl/ojs/index.php/AE2012/article/view/12429/12697

Barron, J. and Noel, S. (2011) 'Valuing soft components in agricultural water management Interventions in meso-scale watersheds: a review and synthesis', *Water Alternatives*, vol. 4, no. 2, pp. 145–155.

Barron, J. and Okwach, G. (2005) 'Run-off water harvesting for dry spell mitigation in maize (zea mays L.): results from on-farm research in semi-arid Kenya', *Agricultural Water Management*, vol. 74, pp. 1–21.

Barry, B., Olaleye A. O., Zougmoré, R. and Fatondji D. (2008) 'Rainwater harvesting technologies in the Sahelian zone of West Africa and the potential for outscaling', IWMI Working Paper 126, International Water Management Institute, Colombo, Sri Lanka, 40 pp.

Bekele, W. and Drake, L. (2003) 'Soil and water conservation decision behavior of subsistence farmers in the Eastern Highlands of Ethiopia: a case study of the Hunde-Lafto area', *Ecological Economics*, vol. 46, pp. 437–451.

Bewket, W. (2007) 'Soil and water conservation intervention with conventional technologies in northwestern highlands of Ethiopia: Acceptance and adoption by farmers', *Land Use Policy*, vol. 24, pp. 404–416.

Boers, T. M. and Ben-Asher J. (1982) 'A review of rainwater harvesting', *Agricultural Water Management*, vol. 5, pp. 145–158.

Bouma, J. A., Biggs, T. W. and Bouwer, L. M. (2011) 'The downstream externalities of harvesting rainwater in semi-arid watersheds: an Indian case study', *Agricultural Water Management*, vol. 98, pp. 1162–1170.

Critchley, W. R. S. (2009) 'Soil and Water Management Techniques in Rainfed Agriculture: State of the Art and Prospects for the Future', Background note prepared for the World Bank, Washington D.C.

Critchley, W. R. S. (2010) *More People, More Trees*, Practical Action, Rugby, UK.

Critchley, W. R. S. and Mutunga, K. (2003) 'Local Innovation in a Global Context: Documenting Farmer Initiatives in Land Husbandry through WOCAT', *Land Degradation & Development*, vol. 14, pp. 143–162.

Critchley, W. R. S., Negi, G. and Brommer, M. (2008) 'Local Innovation in "Green Water" Management', in: D.Bossio and K.Geheb (eds) *Conserving Land, Protecting Water*, CABI Publishing, Oxfordshire, UK.

Critchley, W. R. S., Reij, C. and Seznec, A. (1992) 'Water Harvesting for Plant Production. Volume II: Case Studies and Conclusions for Sub-Saharan Africa', World Bank Technical Paper no. 157, Washington D.C.

Critchley, W. R. S., Reij, C. and Willcocks, T. J. (1994) 'Indigenous soil and water conservation: a review of the state of knowledge and prospects for building on traditions', *Land Degradation & Rehabilitation*, vol. 5, pp. 292–314.

Deressa, T., Hassan, R. M. T. Alemu, Yesuf, M. and Ringler, C. (2009) 'Analyzing the determinants of farmers' choice of adaptation methods and perceptions of climate change in the Nile Basin of Ethiopia', *Global Environmental Change*, vol. 19, no. 2, pp. 248–255.

Drechsel, P., Olaleye, A., Adeoti, A., Thiombiano, L., Barry, B. and Vohland, K. (2010) *Adoption Driver and Constraints of Resource Conservation Technologies in sub-Saharan Africa*, Workshop report, IWMI, West Africa Office, Ghana, FAO, Regional Office for Africa, Ghana and Humboldt University, Berlin, Germany.

Duflo, E., Kremer, M. and Robinson, J. (2011) 'How High Are Rates of Return to Fertilizer? Evidence from Field Experiments in Kenya', *The American Economic Review*, vol. 98, no. 2, pp. 482–488.

Ericksen, P., Thornton, P., Notenbaert, A., Cramer, L., Jones, P. and Herrero M. (2011) *Mapping hotspots of climate change and food insecurity in the global tropics*, CCAFS

Report no. 5. CGIAR Research Program on Climate Change, Agriculture and Food Security (CCAFS), Copenhagen, Denmark.

Feder, G. (1985) 'Adoption of agricultural innovations in developing countries: A survey', *Economic Development and Cultural Change*, vol. 33, no. 2, pp. 255.

Fox, P. and Rockström, J. (2000) 'Water-harvesting for supplementary irrigation of cereal crops to overcome intra-seasonal dry-spells in the Sahel', *Physics and Chemistry of the Earth (B)*, vol. 25, no. 3, pp. 289–296.

Fox, P. and Rockström, J. (2003) 'Supplemental irrigation for dry-spell mitigation of rainfed agriculture in the Sahel', *Agricultural Water Management*, vol. 61, pp. 29–50.

Fox, P., Rockström, J. and Barron, J. (2005) 'Risk analysis and economic viability of water harvesting for supplemental irrigation in semi-arid Burkina Faso and Kenya', *Agricultural Systems*, vol. 83, pp. 231–250.

Hanjra, M. A., Federe, T. and Gutta, D. G. (2009) 'Pathways to breaking the poverty trap in Ethiopia: Investments in agricultural water, education, and markets', *Agricultural Water Management*, vol. 96, pp. 1596–1604.

Hassan, R. and Nhemachena, C. (2008) 'Determinants of African farmers' strategies for adapting to climate change: Multinomial choice analysis', *African Journal of Agriculture and Resource Economics*, vol. 2, no. 1, pp. 83–104.

Hassane, A., Martin, P. and Reij, C. (2000) 'Water harvesting, land rehabilitation and household food security in Niger: IFAD's soil and water conservation project in Illéla District', the International Fund for Agricultural Development (IFAD) and The Programme on Indigenous Soil and Water Conservation in Africa, Phase II (ISWC II), VU University, Amsterdam, the Netherlands, 49 pp.

Hatibu, N., Mutabazi, K., Senkondo, E. M. and Msangi, A. S. K. (2006) 'Economics of rainwater harvesting for crop enterprises in semi-arid areas of East Africa', *Agricultural Water Management*, vol. 80, pp. 74–86.

Hatibu, N., Young, M. D. B., Gowing, J. W. G., Mahoo, H. F. and Mzirai O. B. (2003) 'Developing improved dryland cropping systems for maize in semi-arid Tanazania, Part 1: Experimental evidence for the benefits of rainwater harvesting', *Experimental Agriculture*, vol. 39, pp. 279–292.

Hengsdijk, H., Meijerink G. W. and Mosugu M. E. (2005) 'Modeling the effect of three soil and water conservation practices in Tirgray, Ethiophia', *Agriculture, Ecosystems and Environment*, vol. 105, pp. 29–40.

Hoff, H., Falkenmark, M., Gerten, D., Gordon, L., Karlberg, L. and Rockström, J. (2010) 'Greening the global water system', *Journal of Hydrology*, vol. 384, pp. 177–186.

ISI Web of Knowledge v.5.6. Web of Science, Thomson Reuters, available at: http://apps.webofknowledge.com/WOS

Kabore, D. and Reij, C. (2004) 'The emergence and spreading of an improved traditional soil and water conservation practice in Burkina Faso', EPTD Discussion Paper No. 114, IFPRI, Washington D.C.

Kabubo-Mariara, J. (2007) 'Land conservation and tenure security in Kenya: Boserup's hypothesis revisited', *Ecological Economics*, vol. 64, pp. 25–35.

Kahinda, M., Taigbenu, A. and Boroto, J. (2007) 'Domestic rainwater harvesting to improve water supply in rural South Africa', *Physics and Chemistry of the Earth*, vol. 32, no. 15–18, pp. 1050–1057.

Kato, E., Ringler, C., Yesuf, M. and Bryan, E. (2011) 'Are soil and water conservation technologies a buffer against production risk in the face of climate change? Insights from Ethiopia's Nile Basin', *Agricultural Economics*, vol. 42, no. 5, pp. 593–604.

Kerr, J., Pangare, G. and Lokur Pangare V. (2002) *Watershed development projects in India – an evaluation*, Research report 127, IFPRI, Washington D.C.

Lasage, R., Aerts, J., Mutiso, G.-C. M. and de Vries, A. (2008) 'Potential for community based adaptation to droughts: Sand dams in Kitui, Kenya', *Physics and Chemistry of the Earth*, vol. 33, pp. 67–73.

Liu, B. M., Adebe, Y., McHugh, O. V., Collick, A. S., Gebrekidan, B. and Steenhuis, T. S. (2008) 'Overcoming limited information through participatory watershed management: Case study in Amhara, Ethiopia', *Physics and Chemistry of the Earth*, vol. 33, pp. 13–21.

Maatman, A., Sawadogo, H., Schweigman, C. and Ouedraogo A. (1998) 'Application of zaï and rock bunds in the northwest region of Burkina Faso; Study of its impact on household level by using a stochastic linear programming model', *Netherlands Journal of Agricultural Science*, vol. 46, no. 1, pp. 123–136.

Magombeyi, M. S. and Taigbenu, A.E. (2011) 'An integrated modelling framework to aid smallholder farming system management in the Olifants River Basin, South Africa', *Physics and Chemistry of the Earth, Parts A/B/C*, vol. 36, no. 14–15, pp. 1012–1024.

Mazzucato, V., Niemeijer, D., Stroosnijder, L. and Röling N. (2001) 'Social networks and the dynamics of soil and water conservation in the Sahel', IIED Gatekeeper Series No. 101.

Misselhorn, A. (2005) 'What drives food security in southern Africa? A meta-analysis of household economy studies', *Global Environmental Change*, vol. 15, pp. 33–43.

Moges, G., Hengsdijk, H. and Jansen H. C. (2011) 'Review and quantitative assessment of ex situ household rainwater harvesting systems in Ethiopia', *Agricultural Water Management*, vol. 98, pp. 1215–1227.

Motsi, K. E., Chuma E. and Mukamuri B. B. (2004) 'Rainwater harvesting for sustainable agriculture in communal lands of Zimbabwe', *Physics and Chemistry of the Earth*, vol. 29, pp. 1069–1073.

Mutekwa, V. and Kusangaya, S. (2006) 'Contribution of rainwater harvesting technologies to rural livelihoods in Zimbabwe: the case of Ngundu ward in chivi district', *Water*, vol. 32, pp. 437–444.

NAS (1974) *More Water for Arid Lands*, National Academy of Sciences, Washington D.C.

Pacey, A. and Cullis, A. (1986) *Rainwater Harvesting: The collection of rainfall and run-off in rural areas*, Intermediate Technology Publications, London.

Pachpute, J. S., Tumbo, S. D., Sally, H. and Mul, M. L. (2009) 'Sustainability of rainwater harvesting systems in rural catchment of Sub-Saharan Africa', *Water Resource Management*, vol. 23, pp. 2815–2839.

Reij, C. P. and Smaling, E. M. A. (2008) 'Analyzing successes in agriculture and land management in Sub-Saharan Africa: Is macro-level gloom obscuring positive micro-level change?', *Land Use Policy*, vol. 25, pp. 410–420.

Rockström, J., Karlberg, L., Wani, S.P., Barron, J., Hatibu, N., Oweis, T., Bruggeman, A., Farahani, J. and Qiang Z. (2010) 'Managing water in rainfed agriculture – The need for a paradigm shift', *Agricultural Water Management*, vol. 97, pp. 543–550.

Roose, E., Kabore, V. and Guenat, C. (1999) 'Zai Practice: A West African traditional rehabilitation system for semiarid degraded lands, a case study in Burkina Faso', *Arid Soil Research and Rehabilitation*, vol. 13, no. 4, pp. 343–355.

Sawadogo, R. (2011) 'Using soil and water conservation techniques to rehabilitate degraded lands in northwestern Burkina Faso', *International Journal of Agricultural Sustainability*, vol. 9, no. 1, pp. 120–128.

Schulze, R. (2000) 'Transcending scales of space and time in impact studies of climate and climate change on agrohydrological responses', *Agriculture, Ecosystems & Environment*, vol. 82, no. 1–3, pp. 185–212.

Shiferaw, B. and Holden S. (2001) 'Farm level benefits to investments for mitigating land degradation: empirical evidence from Ethiopia', *Environmental and Development Economics*, vol. 2, pp. 335–358.

Stroosnijder, L. (2009) 'Modifying land management in order to improve efficiency of rainwater use in the African highlands', *Soil & Tillage Research*, vol. 103, pp. 247–256.

Sturdy, J. D., Jewitt, G. P. W. and Lorentz, S. A. (2008) 'Building an understanding of water use innovation adoption processes through farmer-driven experimentation', *Physics & Chemistry of the Earth*, vol. 33, pp. 859–872.

Suri, T. (2011) 'Selection and comparative advantage in technology adoption', *Econometrica*, vol. 79, no. 1, pp. 159–209.

Tabo, R., Bationo, A., Amadou, B., Marchal, D., Lompo, F., Gandah, M., Hassane, O., Diallo, M. K., Ndjeunga, J., Fatondji, D., Gerard, B., Sogodogo, D., Taonda, J.-B. S., Sako, K., Boubacar, S., Abdou, A. and Koala S. (2011) 'Fertilizer microdosing and "warrantage" or inventory credit system to improve food security and farmers' income in West Africa', in: Bationo, A., Waswa, B., Okeyo, J. M., Maina, F. and Kihara, J. M. (eds), *Innovations as Key to the Green Revolution in Africa*, Springer Science + Business, Dordrecht.

Tesfaye, A., Bogale, A., Namara, R. and Bacha, D. (2008) 'The impact of small-scale irrigation on household food security: the case of Filtino and Godino irrigation schemes in Ethiopia', *Irrigation & Drainage Systems*, vol. 22, pp. 145–158.

van Noordwijk, M., Poulsen, J. G. and Ericksen, P. J. (2004) 'Quantifying off-site effects of land use change: filters, flows and fallacies', *Agriculture, Ecosystems & Environment*, vol. 104, no. 1, pp. 19–34.

van Steenbergen, F., Haile, A. M., Alemehayu, T., Alamirew, T. and Geleta Y. (2011) 'Status and potential of spate irrigation in Ethiopia', *Water Resource Management*, vol. 25, pp. 1899–1913.

Vohland, K. and Boubacar, B. (2009) 'A review of in-situ rainwater harvesting practices modifying landscape functions in African drylands', *Agriculture, Ecosystems and Environment*, vol. 131, pp. 119–127.

Walker, T. and Ryan J. (1990) *Village and Household Economies in India's Semi-arid Tropics*, John Hopkins University Press, Baltimore.

Wisser, D. Frolking, S., Douglas, E. M., Fekete, B., Schumann, A. and Vörösmarty, C. (2010) 'The significance of local water resources captured in small reservoirs for crop production – a global-scale analysis', *Journal of Hydrology*, vol. 384, pp. 264–275.

WOCAT, The World Overview of Conservation Approaches and Technologies, www.wocat.net

Zougmorë, R., Mando, A., Stroosnijder, L. and Ouédraogo E. (2004) 'Economic benefits of combining soil and water conservation measures with nutrient management in semi-arid Burkina Faso', *Nutrient Cycling in Agroecosystems*, vol. 70, pp. 261–269.

Chapter 4

Burkina Faso

A cradle of farm-scale technologies

Seraphine Kabore-Sawadogo, Korodjouma Ouattara, Mariam Balima, Issa Ouédraogo, San Traoré, Maurice Savadogo and John Gowing

Introduction

Burkina Faso was a 'laboratory' for water harvesting in the early 1980s. There were several reasons for this. The country had experienced severe drought in the 1970s and the government was leading a popular campaign against desertification. Water harvesting was perceived as having an important role to play, and fortuitously there were traditions of soil and water conservation that lent themselves to improvement. These technologies were well suited to reclaiming barren, compacted land. Furthermore the advent of 'participatory approaches' in the 1980s was taken up with enthusiasm by non-governmental organizations (NGOs) and villagers alike, leading to impressive achievements.

This chapter aims to review experiences of *zaï* (planting pits), stone rows (*cordons pierreux*) and permeable rock dams (*digues filtrantes*) in Burkina Faso with a view to improving the understanding of issues affecting their successful adoption. It provides an update to earlier documentation of these practices (e.g. Critchley et al., 1992; Reij et al., 1996), but differs in that it presents a longitudinal analysis of development initiatives aimed at promoting the adoption of improved water harvesting practices. It attempts to identify the conditions under which successful water harvesting occurs, and to deliver evidence-based policy guidance on how to achieve improved water productivity in dryland agriculture.

Background

Burkina Faso is one of the poorest countries in the world, with a per-capita gross domestic product of US$550 (World Bank 2011). The total population is about 16.5 million (updated from 2007 census) with an annual rate growth of 2.6 per cent. Seventy-three per cent live below the poverty line of US$2 per day. The population is about 75 per cent rural with average density ranging from 25 per km^2 in the east and south to 80 per km^2 on the central plateau. About 90 per cent of the population is engaged in the agricultural sector, and only a small proportion is directly involved in industry and services.

Burkina Faso is a flat country, generally lying between 250 and 400 m above sea level. Soils are often shallow, low in nutrients and vulnerable to erosion by wind and water. They are characterized by their advanced degree of weathering, poor structure, and a low organic matter content. Crusting of sandy clay and sandy loam soils is a common problem which restricts rainfall infiltration. The climate is Sudano-Sahelian and influenced by the movement of the Intertropical Convergence Zone (ITCZ). Rainy seasons generally last from April to October in the south, and from June to September in the north. The annual average rainfall varies from 1000 mm in the south to less than 250 mm in the north. Approximately 25 per cent of the country receives less than 600 mm. Showers are often violent and high intensity, and accompanied by sudden gusts of wind. The average temperature varies seasonally between 27°C and 30°C in the south and between 22°C and 33°C in the north.

Agriculture is mainly at a subsistence level and primarily based on cultivation during the rainy season on small family farms (between 1.5 and 12 hectares (ha) per household) with rudimentary tools and traditional practices. Sorghum (*Sorghum bicolor*), millet (*Pennisetum glaucum*) and maize (*Zea mays*) are the staple foods and are grown on about 80 per cent of the area. Additional cash crops are cotton, groundnuts, cowpeas and yams. The major agricultural constraints include the uneven spatial and temporal distribution of rainfall, the inherently low fertility of soils and low levels of nutrient inputs (Piéri, 1989; Sédogo, 1993; Bationo et al., 1998). Climate-related stress is likely to worsen in the future, as a gradual trend of increased aridity has been observed, with a decrease in the length of the growing season by 20 to 30 days and a southward shift of the 100 mm isohyet (MAHRH, 2006).

In the cotton growing areas (south-west and south-east), maize is planted after cotton to benefit from residual fertilizer. Increased profitability encourages farmers to use improved varieties in the rotation. Because cotton farmers are often equipped with animal traction, they tend to plough their soils before planting. Immigrants from other parts of the country have been settling in the cotton growing area since the 1970s. This creates increasing scarcity of arable land, leading to abandonment of the practice of fallowing as a soil fertility management regime, and the adoption of continuous cropping systems as in the densely populated central plateau region.

There is an estimated irrigation potential of 233,500 ha, but reliable statistics of the area under command are not readily available; however, provisional data from the latest agricultural census (MAHRH, 2009) indicate that there were about 26,000 ha of rice and 34,000 ha of horticultural crops in 2007 – the latter are mainly irrigated from shallow wells, reservoirs and streams. This points to a thriving informal irrigation sector that falls outside the remit of the mainstream government-supported irrigation sector.

There are few perennial rivers and exploitation of groundwater remains limited. The development agenda of the government has prioritized investments in hundreds of small reservoirs that support irrigation, livestock watering, fisheries and domestic water use (Sally et al., 2011). The exact number and condition of these reservoirs and their associated irrigated schemes are not known reliably, but

Cecchi et al. (2009) estimated that there are between 930 and 1400 reservoirs in the country. Such reservoirs are intensively used and generate considerable value. However, current approaches to their management at local level (local water committees, water user associations) are ineffective (Sally et al., 2011).

Given the very limited extent of irrigation coverage, rainfed agriculture dominates – and as a consequence, limiting the impacts of recurring droughts has been high on the development agenda of the government for several decades. Diverse soil and water conservation and water harvesting technologies (stones lines, *zaï*, *demi-lunes*, permeable rock dams, etc.) are used in the central and northern areas of the country (Zougmoré et al., 2004). All regions of Burkina Faso need area-specific technologies for soil moisture and fertility management to cope with climatic variability (Table 4.1).

History of water harvesting practices

External interventions in soil and water conservation in Burkina Faso began in the early 1960s with the 'GERES' project (known by its French acronym) intended to control soil erosion in Yatenga region. Earth bunds were constructed along the contour by machinery over an area of 120,000 ha. Farmers were not involved, and as a result they did not maintain the bunds and indeed often destroyed them (Atampugre, 1993: p. 26). The same approach was taken up 10 years later by a multi-donor funded project (also best known by its French acronym) FDR. Again the level of farmer participation was weak and bunds were often destroyed. An estimated area of 60,000 hectares was treated but within three years most had disappeared (Atampure, 1993: p. 26). One basic problem was that these bunds drained water from the fields, the opposite of what farmers wanted.

The success story of water harvesting in Burkina Faso really began around 1980 and has been well documented (see, e.g. Reij et al., 2009). A change of approach was provoked by the impact of devastating droughts on the densely populated central plateau. Between 1975 and 1985 some villages lost 25 per cent of their population through out-migration. In the early 1980s, groundwater levels dropped by 50 to 100 cm per year, barren land spread and empty, encrusted fields (called *zipele*) stretched across Yatenga. A participatory approach to water harvesting with contour stone lines/bunds (*cordons pierreux*) was tested successfully by the PAF project with support from OXFAM (1979–82) and the technique was then widely promoted over the next 15 years. The success of this intervention (the 'magic stones') became very widely recognized. It was one of the case studies included in the Sub-Saharan Africa Water Harvesting Study (Critchley et al., 1992), where it was reported as a 'water harvesting success story' (see also Critchley, 2010, for a 'before and after' filmed comparison).

Alongside this externally supported improvement, farmers began innovating themselves by improving a traditional practice of planting pits, known locally as *zaï*, in order to reclaim severely degraded land (*zipele*). Ouedraogo and Sawadogo (2001) describe the farmer-led efforts to disseminate the improved

Table 4.1 Constraints and appropriate technologies for soil fertility management in the different agro-ecological zones of Burkina Faso

Rainfall	Constraints	Technologies of soil fertility management	
		Basic technologies	*Complementary technologies*

>900 mm	Low soil fertility
	Soil acidification
	Soil water erosion
	High fertilizer cost
	Continuous tillage
	Land insecurity
	Low technical level
	Lack of credit
		Mineral fertilizer (NPK)
		  and lime
		Organic manure from:
		• Cattle manure or
		  household wastes
		• Compost (cereal
		  straw + household
		  waste)
		• Cover crops
			Annual ploughing
			Ploughing every
			  2 years in the
			  cotton-cereal
			  system
			Ploughing every
			  3 years in the
			  cotton-cereal-
			  legume system.
			Herbicide application.

700–900 mm	Low soil fertility
	Rainfall variability
	High fertilizer cost
	Soil water erosion
	Lack of material for
	  composting
	Continuous tillage
	Land insecurity
	Lack of water
		Mineral fertilizers:
		  NPK+urea
		Organic manure from:
		• Cattle manure and
		  household wastes
		• Compost during the
		  rainy season with
		  addition of Burkina
		  phosphate
		Introduction of legumes
		  into the system
			Yearly ploughing
			Ploughing every
			  2 years in the
			  cotton-cereal
			  system
			- Ploughing every
			  3 years in the
			  cotton-cereal-
			  legume system
			- Herbicide application
			- Strategies against
			  erosion (grass strips,
			  stones lines, earth
			  bunds)

400–700 mm	Insufficient, and poor
	  distribution, of rain
	Water and wind
	  erosion
	High fertilizer costs
	Poor technical level of
	  farmers
	Inadequate equipment
	Grazing pressure on
	  pasture land.
		Mineral fertilizer
		  (NPK + urea)
		Organic manure
		  (or compost)
		Mulching
			Minimum/zero
			  tillage *zaï*;
			Soil and water
			  conservation
			  techniques
			Agroforestry

practice. As summarized in Table 4.2, since the 1980s many donors and NGOs have promoted stone lines and/or *zaï*. These included a major Dutch-funded rural development project (1982–2000) in Kaya region, the German-funded PATECORE project in Bam province and the IFAD-funded PSB project in several provinces (1989–2005). Later projects (Table 4.3) successfully extended the range of interventions. Positive national policies created an enabling

Table 4.2 External interventions in soil and water conservation (see Acronyms and Abbreviations for full names)

Agency	Period	Intervention	Approach
GERES	1962–65	Earth bunds	Top-down, use of machines, no farmer participation
FDR	1972–83	Vegetated earth bunds (with *Andropogon gayanus*)	Village groups, limited farmer participation
PAF/OXFAM	1979–97	Stone lines, use of water-tube levels, *zaï*, agroforestry	Communal and individual
PATECORE	1988–2004	Stone lines, use of water-tube level	Communal and individual
PAE	1981–2000	Stone lines, use of water-tube levels, agroforestry, improved *zaï*	Participatory Voluntary
AFVP	1988–92	Permeable rock dams, stone lines	Communal
PSB Germany	1989–2003	Permeable rock dams, stone lines, *zaï*, *demi-lunes*	Decentralized, communal and individual
PSB Netherlands	1992–2005	Permeable rock dams, stone lines, *zaï*, *demi-lunes*	Decentralized, communal and individual
PSB Denmark	1990–2005	Permeable rock dams, stone lines, *zaï*, *demi-lunes*	Decentralized, communal and individual
CES/AGF	1988–2003	Permeable rock dams, stone lines, *zaï*, *demi-lunes*	Village groups
PEDI	1982–2000	Permeable rock dams, stone lines, *zaï*, *demi-lunes*, agroforestry	Consultation, technical support, loan of equipment
6th FED	1988–93	Earth bunds, stone lines	Village groups
PVNY	1988–93	Stone lines agropastoral integration	Communal

environment from the mid-1980s, and this was recognized as an important factor behind the success (Reij and Steeds, 2003).

By 2001, over 100,000 ha of degraded land had been rehabilitated and the current extent is now estimated at between 200,000 and 300,000 ha with beneficiary households numbering 140,000 to 200,000 (Reij et al., 2009). In some villages, up to 90 per cent of cultivated land has been treated with water harvesting

Table 4.3 Selected water harvesting interventions and associated techniques in Burkina Faso

Project	Funding/supporting organization	Period	Techniques used
GERES-Volta	NEDECO	1962–65	Terraces, earth bunds, diversion ditches, micro-dams
FNGN	FNGN	1967 onwards	Zaï, reforestation, permeable rock dams, earth bunds, compost pits
FDR/FEER	MAHRH	1972 onwards	Vegetated earth bunds, stone lines, grass strips
ADRK	ADRK Kaya	1972–2000	Stone lines, wetland management
PIN	ONG	1978–	Stone lines
PAF	OXFAM	1979–97	Stone lines
PAE	German VSO	1981–2001	Stone lines, compost pits, tree nurseries
PEDI	Netherlands	1982–2000	Stone lines, wetland management
PATECORE	GTZ	1988–2006	Zaï, mulching, stone lines, compost pits, tree nurseries
PAPANAM	CRPA-CN/SPA	1994–99	Stone lines, wetland management
Projet FSA	Ministère de l'Environnement	1994–2000	Demi-lunes, zaï, sub-soiling
PS-CES/AGF	MAHRH	1988–2002	Stone lines, zaï, demi-lunes, permeable rock dams, agroforestry
CES II	INERA/RMARP/ADRK/PEDI	1997–2001	Indigenous and improved soil and water conservation

techniques. Using an indicative yield increase of 400 kg/ha, Reij et al. (2009) estimate an additional harvest of 80,000 tonnes per year. They note, however, that impact has generally not been measured with scientific rigour; many assessments are based on farmers' estimates and others not sufficiently controlled for influencing variables. While numerous research studies have measured positive impacts, most have been limited to one to three years. Nevertheless, they argue, the best indicator of positive impacts is that farmers continue to dig planting pits and construct stone bunds on their own initiative and without external support.

Description of water harvesting technologies

The water harvesting technologies selected for particular analysis here are planting pits (*zaï*), stone lines (*cordons pierreux*) and permeable rock dams (*digues filtrantes*) (see Critchley and Siegert, 1991, for a technical description of these techniques). These are techniques that have survived and evolved as well as spreading widely since the 1980s.

Zaï (planting pit)

This is a traditional technique in parts of Burkina Faso – and elsewhere in the Sahel region (see Chapter 7 on Niger) – which was adapted and improved to recover degraded land. The innovation was first to increase the depth and diameter of the structures, and second to concentrate nutrients and moisture in the pits. As well as aiding water infiltration, they capture windblown soil and organic matter. Manure, compost and mineral fertilizers are usually added to the pits.

Stone lines (cordons pierreux)

Stone lines are barriers along the contour that intercept runoff, thus promoting its even spread and infiltration. Water trickles through the gaps between stones, trapping sediment and organic material upstream of the bund. Before their introduction, much of the manure applied by farmers was washed away during the first rains, but stone lines help to retain it on their fields. Stone lines are built up to 25 cm in height and have a base width of 35 to 40 cm. To increase stability and durability, they are set in a shallow foundation trenches.

Permeable rock dams (digues filtrantes)

These structures are usually built across relatively wide and shallow valleys as a means of controlling gulley erosion, while simultaneously harvesting and spreading runoff. Each dam is typically between 50 and 300 m long, with a maximum height up to 1.5 m. Large stones are used for the outer layer and smaller stones for infill. The dam wall is designed to be semi-permeable, allowing water to permeate slowly through the structure. The crest is level so that if overflow occurs, the whole dam is overtopped evenly. These structures are labour-intensive and usually require mechanized transport for collecting the stone.

Perceptions of key informants[1] on impact of water harvesting practices

Government

The government of Burkina Faso is very supportive of initiatives to improve agricultural production. The current system of land use planning prioritizes sustainable management of natural resources. It recognizes that the overall performance of agriculture in Burkina Faso is highly dependent on an environment that has suffered degradation combined with currently growing population and livestock pressure. In this respect, restoration of soil fertility and the fight against desertification are priorities of the government. The Sustainable

Figure 4.1 Stone lines (W. Critchley)

Agriculture Policy (DPDAD), developed in 1997, identified improved management of soil fertility as essential in achieving the development objective assigned
to the agricultural sector. The government's goal for the agriculture and livestock
sectors is 'to ensure a continuous agricultural production to meet people's needs
while maintaining and improving the quality of life and environment'. Two basic
tools are the National Strategy for Integrated Management of Soil Fertility
(SNGFS) and the Plan of Integrated Soil Fertility (PAGIFS), which were adopted
by the government in 1998.

Extension service

Extension services through various government and non-governmental agencies
play an important role in the process of disseminating water harvesting techniques
in the country. The major problem facing them is funding. To minimize the cost
of extension, the approach has been based around 'farmer innovators' or 'farmer
trainers'. This approach is popular and effective; however, the fact remains that
funding is a serious constraint on wider dissemination of improved practices.

Research

In Burkina Faso, research has been an engine of development of improved techniques for soil and water conservation. Many scientific studies have been carried

Figure 4.2 Permeable rock dam (W. Critchley)

out since the 1970s to optimize water harvesting techniques. The results exist in the form of numerous scientific articles, books, fact sheets and student dissertations. This has positively impacted on yields in the field. However, the researchers note (in common with the extensionists) that funding constraints limit their efforts to further develop these techniques and to measure impacts on yields, soil, the environment, and biodiversity.

NGOs

NGOs perceive many positive impacts of water harvesting, including improvement in the productive capital of farms, a 25 per cent increase in food crop yields, and increased household cash income. Food security and nutrition has improved. Hundreds of thousands of people have benefited from easier access to safe water. The economic situation of women, aided by the proactive strategy of NGOs, has

Figure 4.3 Zaï pit with manure (W. Critchley)

delivered a positive impact on their position and power within communities. Programmes have demonstrated successful techniques in terms of water harvesting, agroforestry, livestock husbandry, village water and microcredit – and these can all now be upscaled. In contrast, the expected positive impacts in terms of building social capital and local capacity did not match the investments made. The management committees of village lands and decentralized public technical services require further strengthening.

Farmers

The conclusion drawn from interviews with villagers is that farmers are convinced of the positive impact of improved water harvesting and conservation techniques. Cereal yield growth is the main indicator according to them. They also perceive positive impact on the rise of the water table, vegetation regeneration, production of fruit trees, the availability of grazing, the management of livestock and recovery of bare land (*zipele*). Women benefit from these developments, particularly the rising water table which reduces time and effort in water collection. They stated that food security in their families has got better; however, they believe their social status has not improved. Elderly women and widows who cultivate alone are particularly disadvantaged. Livestock production has intensified with farmers increasingly adopting a (partial) stall-feeding system which produces a good supply of manure – close to where it can be best used.

Analysis of experience

Evidence of change

Extent of adoption

Recent surveys conducted on a sample of 700 farms where it has been promoted, show that the rate of adoption of water harvesting and associated techniques was just over 50 per cent. Many farmers combine technologies: the most frequent is stone lines combined with *zaï*. The application of organic manure is now widespread, with over 67 per cent of production units having adopted the technology (CILSS, 2009).

In the current situation on the central plateau, there is no village that has not been positively affected in some way. What differentiates them is the duration and level of development (Belemvire et al., 2008). In some villages which have long experience with water harvesting (and associated) technologies, adoption is remarkable: within Zondoma the level of adoption was 94 per cent in Solgomnoré, 83 per cent in Tangaye, 73 per cent in Ranawa and 43 per cent in Salaga; in Bam Province, the extent was 95 per cent in Sankondé; 90 per cent in Rissiam, 72 per cent in Boallé and 42 per cent in Noh; for Yatenga Province, the extent was 81 per cent in Ziga.

Changes in production systems

The agricultural production system is characterized by the predominance of cereal crops (especially sorghum, *Sorghum bicolor* and millet, *Pennisetum glaucum*) and there is an increasing tendency to incorporate improved, high yielding varieties. Investments in the area of soil and water conservation/water harvesting have led to small, but significant, changes in production systems (Belemvire et al., 2008), namely:

- Greater integration of agriculture and livestock allows farmers to reap the benefits of both enterprises. Thus crop residues are collected for feeding draft animals (oxen, horses, donkeys) and for animal fattening (cattle and sheep). Manure is collected, in turn, from livestock to fertilize fields.
- The development of pulse crops is also integral to crop-livestock integration and benefits include the collection of groundnut and cowpea haulms which are used for fattening small ruminants, especially sheep.
- The maintenance of water harvesting structures has led to a change in the agricultural calendar. After the growing season, maintenance of existing bunds needs to be carried out. There are also composting activities that start after harvest. Digging *zaï* and *demi-lunes* is usually accomplished during the offseason.

Impact on crop yields

Studies conducted by teams of researchers have shown that the adoption of water harvesting can double or even triple yields. Many experiments have shown that techniques such as *zaï* and stone lines have an impact on yields of sorghum

(Zombre, 2003; Zougmoré et al., 2004; Sawadogo, 2011; Traoré and Toe, 2008). In general, farmers combine technologies: *zaï* with stone lines; stone lines with grass strips; *zaï* with assisted natural regeneration of on-farm trees. Hullugale et al. (1990) and Maatman et al. (1998) indicate that yields are increased when techniques are combined. Kaboré (2000) found that *zaï* alone increased sorghum yields by 310 kg/ha in the village of Donsin (Namentenga Province), while a combination of *zaï* with stone lines ensured an increase of 710 kg/ha. Farmers, while not expressing impacts in terms of kilograms per hectare, recognize that the techniques have, in general, contributed to the improvement of cereal yields. This has had the effect of improving food security over most of the year: eight to nine months in the event of poor rainfall and the whole year in the case of good rainfall. Indeed, some manage to create a surplus of production several years in succession.

Impact on income

The reduction in grain purchases has stabilized household finances, while surplus production has increased income for some. With improvement in soil fertility, cash crops like cowpeas and sesame have reappeared. Other sources of income have grown with the practice of techniques such as *zaï*: namely hired labour opportunities for digging holes, and the leasing of carts (for manure transport). Surveys conducted in 12 villages by Kaboré and Reij (2003) suggest that new activities help improve farm income by 25–40 per cent. In general, the extra income thus obtained is invested by farmers in the purchase of animals. This is not only a source of savings and liquidity but also future soil fertility through manure production (Ouedraogo et al., 2001). The number of animals per household has increased significantly in the villages since the adoption of water harvesting: 'all this was possible only because there was a surplus of grain'.

Overall, people also recognize a marked improvement in the diversity of herbaceous and woody plants. Perennial grasses are becoming abundant and are maintained in the same fields because of their economic and social value. This is the case with *Andropogon gayanus* – the grass used to vegetate stone bunds, which subsequently colonizes fields if left to do so. This perennial grass has a variety of uses, including crafts.

Impact on migration

The reduction of the rural exodus is another result of increased crop yields: youth out-migration has decreased. Some young people, when interviewed, are reported to have said they did not intend to go elsewhere to seek work because they now have enough to do in the village.

Farmer organizations

An important overall achievement is the growing awareness of detrimental effects of desertification, and the need for farmers to be better organized to fight these effects. The framework in which people work best is that where there is a platform for transferring knowledge and technology between the different stake-holders. Strengthening of village organizations and their better functioning often leads to greater social cohesion. Since 1984 there has been a significant strengthening of organizational and technical skills at the village level (CILSS, 2009).

Changes in water table

Of the 20 villages surveyed by Belemvire et al. (2008), 16 reported a positive impact in terms of a rising water table. CILSS (2009) indicated that in many cases villagers had a unanimous perception of the impact of these water harvesting techniques on the water level in their wells: they emphasized the improved availability of water through better recharging of the aquifer. This considerably reduces the time and efforts involved in raising water, which is particularly beneficial for women. The time-saving has often allowed women to further develop other income generating activities (CILSS, 2009). In some villages the rising water table has led to the creation of vegetable gardens, which has in turn meant increased income for women and improved nutritional quality of meals.

Is the rising water table due to improved rainfall[2] or the spread of water harvesting? Evidence is inconclusive, but there are indications that this rise is at least partially attributable to water harvesting developments since it is reported to be observed in wells that are located close to, or immediately downstream of developments, and not in upstream wells.

Reasons for success

Technologies are based on traditional farming practices and their improvement

Techniques have had a positive impact and become popular because they are based on traditional practices and can be easily mastered and adopted by farmers. This strengthens the confidence of farmers in their own abilities and enables them to undertake maintenance on their own. Thus use of the traditional technique of *zaï* in the early 1980s by Yacouba Sawadogo, a farmer from Gourcy, spread rapidly in Yatenga because of good grain yields obtained by this technique. Subsequently the *zaï* were improved in the early 1980s by a few innovative farmers (Ouedraogo et al., 2001). They managed to gradually expand their fields by rehabilitating degraded lands while developing agro-silvopastoral systems. Their willingness to share their experiences with other farmers led to the creation of a network of innovative farmers. As noted already, the improved *zaï* technique has become a very effective way of rehabilitating severely degraded land. Mechanised *zaï*, made through animal-drawn tools, were then tested and developed by researchers in the 2000s, in order to reduce the drudgery of hand labour.

From 1979 to 1982 the PAF project tested several techniques (*demi-lunes*, stone lines, earth bunds and trash lines) in Yatenga, leaving the villagers the choice of technique (Wright, 1982). The vast majority of farmers opted for stone lines and the project taught them to lay out the contours themselves (Critchley, 2010).

Beginning in 1985, planting of trees alongside bunds was adopted by those who noted the advantages – including bund stabilization and the many uses they could draw from these trees. Species such as *Piliostigma reticulata*, *Acacia nilotica*, *Prosopis* sp, *Ziziphus* sp. and *Azadirachta indica* (Neem) have been planted along stone lines. Nowadays, it is a well-established practice in many places in the central plateau.

CILSS (2009) indicated that the rate of adoption of *zaï* and stone lines was around 53 per cent. The combination of these two techniques is a common practice for most farmers. The application of organic manure is widespread with over 67 per cent of production units having adopted the technology. In the central plateau region farmers are now aware that there are no prospects for agricultural production without resorting to water harvesting.

Control strategies against erosion have evolved from traditional techniques to the concept of managing water, biomass and soil fertility. For rapid adoption and impact, it is important to offer farmers proven solutions which address their specific problems. The range of technologies includes structural techniques for water conservation (*zaï*, *demi-lunes*, stone lines), agronomic measures (mulching, ridging), and agroforestry techniques. All these techniques are simple and can be implemented by farmers.

Recommendations are based on the experiences of previous interventions

Reports of previous, pioneering projects and programmes have provided vital information on the reasons for success. Considerations of the limitations, shortcomings and difficulties in implementation within a project cycle, or between different projects (or successive phases of a project) have contributed to the current successes. Learning from experience has ensured that recommendations have been adapted to the needs of beneficiaries.

Intervention approaches have been adapted and improved

In the original and now discredited top-down approach people 'attended passively to their development', which was determined for them by outsiders. This approach resulted in some cases in neglect of maintenance, mismanagement, destruction of unpopular structures and a low level of technology adoption. The lessons learned have helped to shape a comprehensive approach that aims to engage people, services and projects in the fight against desertification by promoting the active participation of people. Thus the degree of empowerment of farmers and their organizations and the capacity for self-progress has been strengthened.

The approach is based on five principles: integration, collaboration, defining a spatial framework, the creation of an institutional framework and flexible assistance. The participatory approach is a process through which stakeholders share control over development initiatives, decisions and resources that affect them. It involves various actors (land users, NGOs, technical services, etc.) and is exercised at the various levels of rural communities, response agencies and the decentralized structures of government.

This participatory approach is underway in many projects and programmes which work with farmers' associations. It is based on community empowerment. A number of tools and methodological approaches have been developed, to enable a better involvement of communities and stakeholders in the whole process. This process requires the commitment of the population, collaboration with implementing partners, the establishment of an effective organizational framework, consultation with partners and financial support.

Training is provided in execution of works

It was NGOs who first provided training of farmers in courses carried out from 1977 to 1982 by the Institute for Research and Training, Education and Development (IFERD), and afterwards by the International Centre for Education and Concerted Development (CIEPAC). They introduced and distributed the 'water-tube level' and trained field managers and farmers to use it. The Agroforestry Project (PAF) in Yatenga was the first that focused on farmers who had begun to organize themselves to improve their own fields.

Efforts promoted the production and use of organic matter

The majority of soils in Burkina Faso are characterized by low organic matter content (Lompo ct al., 1994). Good agricultural practices therefore require production and increased use of organic matter. The use of organic fertilizers is a key element of sustainable intensification of agriculture. Work done between 1970 and 1990 (AG, 1996) to characterize the different types of organic matter and available production technology produced information sheets that provide guidance on the value of organic manure and management practices for its use on the farm. Diversification of production techniques that take into account the variability of socio-economic conditions of farmers, such as those tested and promoted by INERA (amongst others), have succeeded. Examples are composting in the rainy season (Ouattara et al., 2005), and the techniques of producing organic matter from crop residues. The CEAS system promotes the use of two compost pits for timely production and four pits for sustained production (CEAS, 2004). Other improved techniques of improving organic matter have been reviewed by Traore and Toé (2008).

The increased availability of organic manure to the majority of family farms in northern Burkina Faso, has led to improved crop yields and soil fertility. Belemviré et al. (2003, 2008) indicate that improved methods of manure and compost making lead to the best grain yields; these technologies have been spread by training farmers. Widespread adoption has contributed to greater integration of crops and livestock for the production of organic manure and its use in *zaï*. Various government programmes have supported these efforts. National days for farmers have contributed to promoting the production and use of organic fertilizer, with over 1.5 million compost pits made during the period from 2001 to 2006 (Traore and Toé, 2008).

Challenges and constraints

While the impact on crop yields and soil fertility of water harvesting and associated technologies have been scientifically proven (see Chapter 2 for the concept of 'water harvesting plus' or WH+) the adoption rate remains below projections. In most situations this has been related to the availability of equipment for implementation, the heaviness of work in terms of labour, lack of funding and insecurity of tenure (Traoré and Toé, 2008).

Labour requirement

The construction of contour stone lines requires labour for collection, transport and alignment. Kaboré (1993) indicates that the number of hours varies but is in the order of 100 hours per ha on a typical family farm. In some circumstances it can reach 600–700 hours per ha when stones supplies are distant. The contribution of women and children may be substantial not only for collection but also for the alignment of the stone lines. Another major constraint is the availability of transport for stones. Even with a donkey cart, the job remains difficult. The amount of stone needed to complete development of a 1-ha plot is close to 40 tonnes for 300 m of stone lines.

As for the workload required for *zaï*, it comprises labour for digging of the pits, and for production of manure and its transport and spreading into the pits. More than 900 hours are required per hectare of which 600 may be for digging. This explains the modest incremental area that can be developed each year: typically 0.25 hectares per household.

However poorer families may benefit from donor-supported interventions that address extents of land that cover the fields of multiple households, with project supported construction of stone lines or permeable rock dams (Haggblade and Hazell, 2009).

Insufficient organic matter

Zaï demand more than labour; there is also the requirement for organic manure which is not always available in sufficient quantity. It follows then there is a need for stall-fed animals (at least in part) and/or making of compost pits to produce organic manure. Given the quantities of organic matter required per hectare (more than 5 tonnes), producers struggle to produce this amount.

Lack of equipment

For *zaï* and *demi-lunes*, equipment is required to produce and transport manure. As for stone lines and permeable rock dams, the transportation of stone is out of the reach of farmers when it comes to great distances. Lack of equipment is therefore one of the reasons for low rates of adoption or non-achievement of expected results.

Weakness of farmer organizations

The problem of representativeness of village groups and therefore the participation of producers is an obstacle in some communities. Indeed, farmers organized in groups are often the most influential and others may feel excluded. The question arises whether all households have access to these groups. Poor households may be marginalized and excluded. These are likely to be households made up of old people and widows.

Difficulties related to land tenure

Land is owned communally but managed by individual households. There is no clear evidence that the prevailing land tenure system interferes seriously with the development of water harvesting facilities. However, land law is unfavourable to women since they can usually get access to a plot with the agreement of the husband.

Inappropriate approaches

The failure of early interventions such as GERES can be explained by the top-down implementation approach and little consideration given to the contextual appropriateness of the techniques. Later projects learned and adopted a more participatory approach.

Knowledge transfer constraints

Over and above the reasons mentioned above, the high level of illiteracy limits dissemination of good practice.

Conclusions

Given the mass poverty which Burkina Faso still faces today, the quest of the country remains economic and social development. This is understood as sustained growth in average income, the satisfaction of basic needs, poverty reduction and improvement of human capacity. The challenge of sustainable development is that which combines economic efficiency, social equity and conservation of the environment. Water harvesting technologies developed through research and adaptation of local knowledge have contributed to sustainable intensification of agriculture under very challenging conditions over the past 30 years. The advent of 'participatory approaches' in the 1980s was taken up with enthusiasm by NGOs and villagers alike leading to impressive achievements. A farmer-managed agro-environmental transformation has occurred on the central plateau. This can be seen as a 'bright spot' where poor farmers have enhanced their food security while responding to climate change. Despite the continuing challenges, Burkina Faso truly represents a cradle of farmer-scale water harvesting technologies that have matured impressively over the last 30 years.

Notes

1 Data collected by the authors during fieldwork under the WHaTeR project in 2011.
2 In Burkina Faso (as represented by Ouahigouya on the central plateau), there has been a marked recovery from the droughts of the early 1970s, and a gradual improvement in rainfall through the 1980s (annual average of 532 mm) to the 1990s (ann. av. 664 mm) and into the 2000s (ann. av. 689 mm). Source: Kabore, S. (pers. comm.) in Critchley (2010).

References

Atampure, N. (1993) *Behind the Lines of Stone*, Oxfam Publication, Oxford.
Bationo, A., Lompo, F. and Koala, S. (1998) 'Research on nutrient flows and balances in West Africa: state-of-the-art', *Agriculture, Ecosystems & Environment*, vol. 71, pp. 19–35.
Belemvire, A., Maiga, H., Sawadogo, M., Savadogo, M. and Ouedraogo, S. (2008) 'Evaluation des impacts biophysiques et socio-économiques des investissements dans les actions de gestion des resources naturelles au Nord du Plateau Central du Burkina Faso', Rapport de Synthese Etude Sahel Burkina Faso, Ouagadougou, Burkina Faso: Comité Inter-Etats pour la Lutte contre la Secheresse dans le Sahel (CILSS) and Vrije University Amsterdam.
Cecchi, P., Meunier Nikiema, A., Nicolas, M. and Sanou, B. (2009) 'Towards an atlas of lakes and reservoirs in Burkina Faso', In Small Reservoirs Toolkit, available at: www.smallreservoirs.org/full/toolkit/docs/IIa%2002%20Faso%20MAB_ML.pdf
Critchley, W. (2010) *More People More Trees: Environmental recovery in Africa*. Practical Action Publishing, Rugby.
Critchley, W., Reij, C. and Seznec, A. (1992) 'Water harvesting for plant production', World Bank Technical Paper Number 157, World Bank, Washington D.C.
Critchley, W. and Siegert, K. (1991) *Water Harvesting*. FAO, Rome.
Haggblade, S. and Hazell, P. (eds) (2009) *Successes in African agriculture: lessons for the future*, Johns Hopkins University Press, Baltimore, MD.
Hien, F. and Ouédraogo, A. (2001) 'Joint analysis of the sustainability of a local SWC technique in Burkina Faso', in: C. Reij and A. Waters-Bayer (eds), *Farmer innovation in Africa: a source of inspiration for agricultural development*, Earthscan Publications Ltd, London.
Hulugalle, N. R., De Koning, J. and Matlon, P. J. (1990) 'Effect of rock bunds and tied ridges on soil water content and soil properties in the Sudan savannah of Burkina Faso', *Trop. Agric. (Trinidad)*, Vol. 67, pp. 149–153.
Kaboré, P. D. (2000) 'Performance des technologies de conservation des eaux et du sol en champs paysans a Donsin, Burkina Faso', *Annales de l'Universite de Ouagadougou Serie*, Vol. 13.
Kaboré, D. and Reij, C. (2003) 'The emergence and spread of an improved traditional soil and water conservation practice in Burkina Faso', Conference Paper no. 10, pp. 22,

Paper presented at the InWEnt, IFPRI, NEPAD, CTA conference 'Successes in African Agriculture'; Pretoria, December 1–3, 2003, South Africa.

Maatman, A., Sawadogo, H., Schweigman, C. and Ouedraogo, A. (1998) 'Application of zaï and rock bunds in the northwest region of Burkina Faso; Study of its impact on household level by using a stochastic linear programming model'. *Netherlands Journal of Agricultural Science*, vol. 46, no. 1, pp. 1–10.

MAHRH (2006) 'Politique nationale de développement durable de l'agriculture irriguée, Stratégie, plan d'action, plan d'investissement à l'horizon 2015', Rapport principal, MAHRH, Ouagadougou, Burkina Faso.

MAHRH (2009) 'Résultats de la phase 1 du recensement général de l'agriculture: Volet cultures irriguées', Rapport Provisoire, Direction Générale des Prévisions et des Statistiques Agricoles, Ouagadougou, Burkina Faso.

Ouédraogo, M., Ouédraogo, A., Kaboré, D., Bemviré, A., Zida, C. and Reij, C. (2001) *Etude d'impact des actions de CES, d'agroforesterie et d'intensification agricole au plateau central*, Burkina Faso.

Ouédraogo, A. and Sawadogo, H. (2001) 'Three models of extension by farmers innovators in Burkina Faso', in: C. Reij and A. Waters-Bayer (eds), *Farmer Innovation in Africa: a source of inspiration for agricultural development*, Earthscan, London.

Piéri, C. (1989) *Fertilité des terres des savanes. Bilan de trente ans de recherche et de développement agricoles au sud du Sahara*, CIRAD, Paris.

Reij, C. and Steeds, D. (2003) 'Success stories in Africa's Drylands: supporting advocates and answering skeptics', A paper commissioned by the Global Mechanism of the Convention to Combat Desertification, Vrije University and Centre for International Cooperation, Amsterdam, Netherlands.

Reij, C., Tappan, G. and Smale, M. (2009) 'Agro-environmental transformation in the Sahel', IFPRI Discussion Paper 00914, International Food Policy Research Institute.

Sally, H., Lévite, H. and Cour, J. (2011) 'Local water management of small reservoirs: Lessons from two case studies in Burkina Faso', *Water Alternatives*, vol. 4, no. 3, pp. 365–382.

Sawadogo, H. (2011) 'Using soil and water conservation techniques to rehabilitate degraded lands in northwestern Burkina Faso', *International Journal of Agricultural Sustainability*, vol. 9, no. 1, pp. 120–128.

Sédogo, M. P. (1993) 'Evolution des sols ferrugineux lessivés sous culture: Incidence des modes de gestion sur la fertilité', Université Nationale de Côte-d'Ivoire.

Traoré, O., Somé, N. A., Traoré, K. and Somda, K. (2007) 'Effect of land use change on some important soil properties in cotton-based farming system in Burkina Faso'. *Int. J. Biol. Chem. Sci.*, vol. 1, no. 1, pp. 7–14.

Traoré, K. and Toé, A. M. (2008) 'Capitalisation des initiatives sur les bonnes pratiques agricoles au Burkina Faso'. *Rapport de consultation*, MAHRH/DVRD, Ouagadougou, Burkina Faso, 99 p.

World Bank (2011) Burkina Faso: country brief. http://go.worldbank.org/HFJD4UQ0M0

Zombre, N. P. (2003) 'Les sols tres degrades (Zipella) du Centre Nord du Burkina Faso: Dynamique, Caracteristiques morpho-bio-pedologiques et impacts des techniques de restauration', *These de Doctorat des Sciences Naturelles*, Universite de Ouagadougou, Burkina Faso.

Zougmoré, R., Ouattara K., Mando, A. and Ouattara, B. (2004) 'Rôle des nutriments dans le succès des techniques de conservation des eaux et des sols (cordons pierreux, bandes enherbées, zaï et demi lunes) au Burkina Faso', *Science et changements planétaires/Sécheresse*, vol. 15, no. 1, pp. 41–48.

Chapter 5

Ethiopia

Opportunities for building on tradition – time for action

*Adane Abebe, Ralph Lasage, Ermias Alemu,
John Gowing and Kifle Woldearegay*

Introduction

Ethiopia is one of the countries in Sub-Saharan Africa most seriously affected by
land degradation, which is a major cause of the country's low and declining agricultural productivity, persistent food insecurity, and rural poverty. There is a close
relationship between land degradation, drought, crop failure and malnutrition. The
annual cost of land degradation is estimated to be two to three per cent of agricultural GDP. This is a significant loss for a country where agriculture accounts for
nearly 50 per cent of GDP, 90 per cent of export revenue, and is a source of livelihood for more than 80 per cent of the country's 82 million people. Limited adoption of effective soil and water conservation practices and the breakdown of
traditional land productivity restoration measures contribute to land degradation,
which represents an acute challenge to rural livelihoods. Addressing this problem
has been consistently identified as a major priority in national strategies and policy
documents. However, existence of and potential for wider adoption of water
harvesting practices has been largely neglected with agricultural water management being viewed as synonymous with irrigation. Recent adoption of the 'Plan for
Accelerated and Sustained Development to End Poverty' (PASDEP) strategy to
reverse land degradation places emphasis on community-based approaches to
watershed management, and appears to mark a change in the perception of the
importance of soil and water harvesting practices. This chapter reflects on experiences in Ethiopia with water harvesting techniques and identifies the importance of
their wider promotion as part of future strategy for sustainable land management.

Ethiopia: background[1]

Ethiopia is a country of great diversity. Topography encompasses high and
rugged mountains, flat-topped plateaux and deep gorges; with altitudes ranging
from 110 m below sea level in the Denakil Depression to over 4600 m above sea
level in the Simien Mountains. There are diverse soil types in the country.
Cambisols are predominant over much of the highlands, while Vertisols occur in
large areas of the south-eastern highlands and in the south-western parts of the

country. Regosols occupy much of the Somali Plateau and the north-eastern part of the country. Ethiopia has a tropical monsoon climate, but with wide topographic-induced variation. Three climatic zones can be distinguished: a cool zone consisting of the central parts of the western and eastern section of the high plateaux, a temperate zone between 1500 and 2400 m above sea level, and the hot lowlands below 1500 m. Mean annual potential evapotranspiration varies between 1700 and 2600 mm in arid and semi-arid areas and 1600 and 2100 mm in dry sub-humid areas. Average annual rainfall varies from over 2000 mm in some areas in the south-west to less than 100 mm in the Afar Lowlands of the north-east. Rainfall is highly erratic, and falls often as convective storms, with very high-intensity and extreme spatial and temporal variability. The result is that there is a high risk of annual droughts and intra-seasonal dry spells. Water harvesting appears to have the potential to reduce the impacts of these risks on the local populations.

Considering the water balance and the length of the growing period, Ethiopia can be divided into three major agroclimatic zones:

- areas without a significant growing period, with little or no rainfall (eastern, north-eastern, south-eastern, southern and northern lowlands);
- areas with a single growing period and one rainy season from February/March to October/November, covering the western half of the country, with the duration of the wet period decreasing from south to north; and
- areas with a double growing period and two rainy seasons (*Belg* and *Meher*) which are of two types: bimodal type 1 and bimodal type 2. The region of type 1 in the east of the country has a small rainfall peak in April and a major one in August. The region of type 2, covering most of the lowlands of the south and south-east, has two distinct wet periods, from February to April and from June to September, interrupted by two clear-cut dry periods. The peak rainfall months are April and September.

The country's total population in 2012 was 82 million (extrapolated from 2007 census), of which about 80 per cent is rural and is dependent on agriculture. Food insecurity is persistent and even during good years the survival of some 4–6 million people depends on international food assistance.

The following main agricultural production systems can be distinguished:

- The highland mixed farming system is characterized by a very low level of specialization of production based on environmental and land suitability and is practised by about 80 per cent of the country's population on about 45 per cent of the total land mass in areas at more than 1500 m above sea level. Livestock production is an integral part of the system, but is increasingly being restricted to stall feeding of animals due to scarcity of land. Declining landholding sizes because of population growth and deteriorating soil fertility are among the biggest challenges facing this production system.

- The lowland mixed agricultural production system is practised in low-lying plains, valleys and mountain foothills, which include the northern parts of the Awash and the Rift Valley with elevations of less than 1500 m. These areas mainly produce drought-adapted varieties of maize, sorghum, wheat and teff (*Eragrostis tef*), along with some oil crops and lowland pulses.
- The pastoral complex supports the livelihood of 10 per cent of the total population living in the Afar and Somali Regions and the Borena Zone. Livestock is the major livelihood basis of these populations who are highly mobile in search of water and grazing. Some lowland varieties of maize, sorghum and other cereals are also cultivated on flood plains or as rainfed crops.
- Shifting cultivation is practised in the southern and western part of the country. Fields are usually left fallow after short periods of cultivation to re-vegetate (usually one to two years). Clearing of the vegetation cover is carried out through burning during the dry seasons before the planting of sorghum, millet, sesame, cotton and ginger. These areas have low population densities and in some of them, livestock production is constrained by tsetse fly infestation.
- Commercial agriculture has only emerged very recently. Access to land and infrastructure-related problems as well as investment insecurity have hindered the growth of this system of production.

The surface water resource potential is considerable, but little developed. Most of the rivers in Ethiopia are seasonal and about 70 per cent of the total runoff is obtained during the period June to August. The groundwater potential of the country is not known with any certainty, but so far only a small fraction of the groundwater has been developed and this is mainly for local water supply purposes. Traditional shallow wells are widely used by nomads for livestock watering.

Irrigation in Ethiopia dates back several centuries, if not millennia, while 'modern' irrigation was started in the early 1950s for commercial scale sugar cane production. The area equipped for irrigation is around 300,000 hectares (ha), which is only 11 per cent of the irrigation potential. About 62 per cent of the area equipped for irrigation is located in the Rift Valley, while 29 per cent is found in the Nile basin. The remaining 9 per cent is located in the Shebelli-Juba basin. Though quantitative information is not available, spate irrigation and flood recession cropping are practised in the lowland areas of the country, particularly in Dire Dawa, Somali, East Amhara and Tigray in the eastern and north-eastern parts of the country.

Four categories of irrigation schemes can be distinguished:

- Traditional irrigation schemes: These schemes are constructed by farmers on their own initiative and vary from less than 1 ha to 100 ha. The total irrigated area is estimated to be about 138,000 ha and about 572,000 farmers are involved. Traditional water committees administer the water distribution and coordinate the maintenance activities of these schemes.
- Modern small-scale irrigation schemes: These schemes are constructed by the government/NGOs with farmer participation for irrigating up to 200 ha.

They are generally based on direct river diversions but they may also involve micro-dams for storage. The operation and maintenance of the schemes are the responsibility of the water users, supported by the regional authorities/bureaus in charge of irrigation development and management. Water Users Associations (WUAs) have been formally established in some schemes but traditional water management dominates in most of these schemes.

- Modern private irrigation: Private investment in irrigation has recently re-emerged with the adoption of a market-based economy policy in the early 1980s. Virtually all irrigated state farms were privately owned farms until nationalization of the private property in the mid 1970s.
- Public irrigation schemes: These schemes comprise medium- and large-scale irrigation schemes with areas of 200 to 3000 ha and above 3000 ha respectively and a total estimated area of about 97,700 ha. They are constructed, owned and operated by public enterprises. These schemes are concentrated along the Awash River and were constructed in the 1960s and 1970s as either private farms or joint ventures.

Both irrigated and rainfed agriculture are important in the Ethiopian economy. Virtually all food crops in Ethiopia come from rainfed agriculture with the irrigation sub-sector accounting for only about 3 per cent of the food crops. Export crops such as coffee, oilseed and pulses are also mostly rainfed, but industrial crops such as sugar cane, cotton and fruit are irrigated. Other important irrigated crops include vegetables and fruit trees in medium- and large-scale schemes and maize, wheat, vegetables, potatoes, sweet potatoes and bananas in small-scale schemes.

Even in good years Ethiopia cannot meet its large food deficit through current rainfed production. Growing population pressure in the highland areas of rainfed agriculture together with recurrent drought and the acute problem of land degradation and perceived under-utilization of abundant water resources has secured irrigated agriculture a prominent position on the country's development agenda. However, Ethiopia will become a physically water scarce country by the year 2020 (Awulachew et al., 2005) and competition for water resources will limit scope for expansion of irrigation. Water harvesting for agriculture has not received high priority but its present extent and future potential will become apparent in the discussion which follows.

History and importance of soil and water conservation and water harvesting practices

Perhaps the most notable example of traditional soil and water conservation technologies, including water harvesting, can be seen in the Konso terrace systems of the Southern Nations, Nationalities and Peoples Region (SNNPR). The Konso have developed a combination of soil and water conservation structures, including stone terraces, water harvesting, square-ridged basins, manuring, intercropping and agroforestry to obtain food from land they cultivate permanently despite low and unpredictable rainfall. The long tradition and strong connection between landscape and culture (Watson, 2009) resulted in the Konso terrace system achieving UNESCO world heritage status in 2011 (UNESCO, 2011). Evidence of indigenous knowledge from other regions of Ethiopia has also been documented (Asrat et al., 1996; Abay et al., 1999; Michael and Herweg, 2000; Dixon, 2001).

Since the 1973–74 famine, various government agencies and NGOs have established projects to promote adoption of water harvesting. Some 'bright spots' have been documented as evidence of success (Mintesinot et al., 2005) but many initiatives were less successful. Recognizing the importance of first understanding indigenous practices, Kruger et al. (1996) reported an attempt to create an inventory of indigenous soil and water conservation techniques in representative parts of Ethiopia. In 2007, the Ministry of Agriculture and Rural Development established the Country Partnership Programme for Sustainable Land Management in Ethiopia (CPPSLM). The approach starts with collective learning about problems and possible solutions, moves on to the description and evaluation of local experiences and finally to the joint selection of potential solutions. This initiative has compiled a comprehensive database of 52 known technologies and 28 approaches common in Ethiopia. In 2010 an overview was published by the Ethiopian Overview of Conservation Approaches and Technologies (EthiOCAT) network containing WOCAT descriptions of 35 techniques and eight approaches (Dale, 2010). In view of the very limited development of groundwater it is not surprising that the knowledge captured here is in stark contrast with much of the experience in India where there has been a much greater emphasis on groundwater recharge. A notable exception is reported in Box 5.1.

In order to understand the generally limited impact of attempts to promote water harvesting and sustainable land management it is important to consider the policy environment. In recent history the governance of Ethiopia has changed several times with consequent changes in policy affecting agricultural development. Before 1972, the country was essentially feudal, with no special attention to soil and water conservation although traditional water harvesting techniques were evident. Between 1972 and 1991 the country was under the Derg regime which promoted state ownership and top-down development planning including initiatives affecting soil and water conservation. From 1991 to 1993 a transitional government was in place and since 1993 Ethiopia has been a parliamentary democracy. The key transition dates of 1972 and 1993 were adopted as the framework for investigating the timeline of soil and water conservation and water harvesting promotion and adoption in Ethiopia (Lasage et al., 2011).

Interventions started as a response to the 1973–74 drought and famine with the introduction of food-for-work (FFW) programmes which were intended to generate employment opportunities to the people affected by the drought. Water harvesting activities included construction of ponds, micro-dams, bunds and

Box 5.1   Groundwater recharge through water harvesting: the case of Abreha Weatsbeha

Abreha Weatsbeha, located about 55 km north of Mekelle in Tigray, Northern Ethiopia, used to be one of the most food insecure areas in the region. The watershed covers 28.2 km² and its topography ranges from mountain slopes to a flat and wide valley floor. The average annual rainfall is around 650 mm, and land use varies, with half of the area under annual crops. To address its problems of severe land degradation, a mixture of soil and water conservation interventions were carried out through cooperation between communities, local authorities and the regional government. The interventions began ten years ago and include stone check-dams, percolation ponds (pictured), deep trenches, stone/soil bunds, area closures (total protection from grazing), as well as afforestation in the higher reaches of the watershed. One of the most remarkable impacts of these interventions has been recharge of the groundwater. This has allowed an expansion in irrigation – through the construction of around 600 hand-dug shallow wells. Another very visible effect has been natural regeneration of the indigenous *Faidherbia albida*, which has the well-deserved reputation of being the best agroforestry tree in Africa's drylands.

Photo 5.1a Percolation ponds harvesting water from the hillside to recharge groundwater (K. Woldearegay)

Despite the considerable challenges faced, through integrated and community-owned watershed management, experience at Abreha Watsbeha shows that it is possible to ensure food security while simultaneously creating a healthy and resilient natural environment.

Photo 5.1b Natural regeneration of *Faidherbia albida* trees on the valley floor (W. Critchley)

terraces in most parts of the drought affected areas in Tigray, Wollo and Hararghe provinces (Kebede, 1995). The policy of rural transformation adopted by the Derg regime was based upon state ownership of land with centralised planning and top-down development initiatives. Since the new government took over in 1991, there has been a gradual shift to more participatory community-driven approaches. This is, for example, reflected in changes to the approach adopted by the Managing Environmental Resources to Enable Transition (MERET) programme (see below).

The National Environmental Policy of Ethiopia, which was adopted in 1997 contains two goals that are important to water harvesting projects (EPA, 1997):

- To involve water resource users, particularly women and animal herders, in the planning, design, implementation and follow up in their localities of water policies, programmes and projects so as to carry them out without affecting the ecological balance.
- To promote, to the extent possible, viable measures to artificially recharge ground and surface water resources.

The policy mentions 'Developing small-scale irrigation and water harvesting schemes in arid, semi-arid, and dry, sub-humid areas', as one of three adaptation options for the semi-arid lowlands of Ethiopia (FDRE, 2001).

Awulachew et al. (2005) reviewed recent experiences and future opportunities for promoting small-scale irrigation and water harvesting in Ethiopia. They revealed mixed perceptions about the impacts of past initiatives. Micro-dams were particularly noted for negative health and environmental impacts in all the regions covered in the study. There was a general perception in all regions that the current trend of low performance is related to flawed project design and lack of adequate community consultation during project planning. They noted a general lack of social and economic research on costs and benefits of interventions in a multiple livelihoods perspective (see Box 5.2 for an exception). Returns on investments and how they compare to alternative options of livelihood generation, impacts on poverty, incomes, and equity, are critical issues that need attention. In most cases, critical issues constraining viable production and poverty impacts on existing schemes are not clearly understood. Issues include property rights and access of the poor to land and water, as well as access to input and output markets.

Key water harvesting development practices: the technologies and the approaches

Within the context of the current enquiry the focus is on two regions of southern Ethiopia (SNNPR and Oromia) and the four key water harvesting interventions that were identified in those regions. These are: houschold ponds, sand dams, spate irrigation and the integrated micro-watershed (MERET) approach. Descriptions of these techniques and their presence in Ethiopia are given in Table 5.1. Available evidence suggests that there are success stories but the

---

Box 5.2 Survey of impacts of water harvesting on livelihoods in Ethiopia (Ayele et al., 2006)

A survey was administered in 2033 households selected from four regions (Amhara, Oromia, SNNPR and Tigray). Across all regions the dominant water harvesting technology was the household pond system (see below). The total cost of construction was Birr 6000 to 8000 (at the 2012 exchange rate 100 Birr = US$5.6). Supplementary irrigation for horticultural crops was shown to deliver a positive economic return while full irrigation generally showed a negative return. When analysed from a livelihood perspective the impact was shown to be more positive; the vast majority of farmers reporting increased food security, increased cash income and improved diet.

Table 5.1 Description of selected water harvesting and soil and water
conservation interventions

Pond	Hand dug open reservoir to store water collected from local catchment. Seepage losses can be reduced by using lining (concrete or plastic). Sizes vary from 30 m³ (individual household use) to 20,000 m³ (community use). Ponds are normally constructed in non-sloping or slightly sloping terrain, and store surface runoff. These are simple structures that can be constructed by untrained labourers. When lining is used, some expertise is necessary. Water is extracted using a bucket or foot pump. A roof may be constructed to reduce evaporation losses; then more correctly know as a cistern (Fox et al., 2005; Critchley, 2009). Implemented in Afar, Amhara, Oromia, SNNP, Somali, and Tigray regions.
Sand dam	Impermeable concrete or stone masonry structure constructed across a seasonal river. Increases water storage capacity by enlarging the aquifer within the original river bed, through accumulation of sand and gravel particles against the dam. The sub-surface reservoir is recharged during flash floods and when the reservoir is filled surplus water passes over the dam. The stored water is captured for use through digging a scooping hole, or constructing an ordinary well or tube well. By storing the water in the sand, it is protected against high evaporation losses and contamination (Lasage et al., 2008). Implemented in Oromia, SNNP and Somali regions
Floodwater harvesting	Diversion of floodwater (spate flows) from beds of ephemeral rivers where semi-arid mountain catchments border lowlands, through open intakes, by diversion spurs or by bunds built across the river bed to spread over large areas as irrigation water and to be partially stored in the soil to cultivate crops, feed drinking water ponds, or irrigate pasture areas. Dry-planting is carried out. The intake structure often needs to be rebuilt after a flood. Normally communal activity requiring social organization. (Tesfai and de Graaff, 2000; van Steenbergen et al., 2010; van Steenbergen et al., 2011). Spate irrigation is on the increase in the semi-arid parts of the country: Tigray (Raja, Waja), Oromia (Bale, Arsi, West and East Haraghe), Dire Dawa Administrative Region, SNNP (Konso), Afar and Amhara (Kobe) regions.
MERET	The MERET (Managing Environmental Resources to Enable Transition) project, started in the 1980s as a food-for-work (FFW) programme implementing soil and water conservation measures on micro-watersheds, supported by WFP. It evolved into a community-based and people centred participatory catchment approach to soil and water conservation, where WH for income generating activities at household level is seen as an entry point. Catchment management activities (focus on restoration of degraded land) include soil and water conservation and water harvesting on communal land. Implemented in all regions. It is aimed at improving food security and livelihoods of chronically food insecure and impoverished communities, by providing food-for-work incentives to enable local people to invest in land management (and water harvesting practices. Over the years, MERET has covered more than 600 sub-watersheds each with 300 to 2000 participating households, in 74 *woredas* (districts) in six regions (including SNNPR and Oromia) and has rehabilitated over 400,000 ha of heavily degraded lands.

adoption of water harvesting is in all cases mediated by external intervention by government agencies or NGOs and has not occurred spontaneously by individual farmers or communities.

Pond systems

Household ponds are open hand-dug reservoirs of 60–150 m^3 designed to store water in the rainy season. Near Konso and Alaba, both located in SNNPR, thousands of ponds have been constructed, mainly during the national programme on construction of water harvesting structures. The stored water is used for multiple purposes, some households use it for drinking water when other sources are lacking, many households use it for watering livestock, and all households use it for supplemental irrigation.

There are large differences between projects with regard to what part of the costs is covered by the household and what is covered by a donor or the government. This varies from 100 per cent of all costs covered by the household, to 100 per cent of all costs covered by the donor. Lining adds considerably to total costs and is generally only used when it is provided by an external agency.

Pond maintenance consists of desilting the pond and silt trap and takes approximately five days per year and these costs are covered by the household. Repairs for the cement or geomembrane lining have to be done by experts. The availability of water near the house reduces walking time for watering livestock. Irrigation of crops is said to be done by all members of the household; the time spent on irrigation cannot be spent on other activities.

Economic analyses indicate that for many farmers the long payback periods in relation to high investments at the start deters them from adopting the technology without financial support (Shiferaw and Holden, 2001; Fox et al., 2005; Dawa, 2006). Without external subsidies most households will probably not construct ponds. For the pond programmes a strong external promoter was present, but estimates suggest that only 25 per cent of ponds continue to function after a few years. There is a need for better impact studies.

Sand dams

Sand dams were successfully constructed in the east of Ethiopia by government organizations in the 1970s, but the technique was not introduced in other parts of the country. In 2006 an international team of organizations started a project in the Borana region, introducing the sand dam technique based on prior experiences in Kenya. After the dams proved to be applicable in that region and increasing water availability, they were introduced to other regions in Ethiopia (SNNPR and Somali) through cooperation with Ethiopian NGOs.

For sand dams it seems that for a long time a promoter was missing, hampering the replication of a suitable technique for Ethiopia in similar hydro-ecological zones. Since the arrival of a consortium working on sand dams, the technique has spread to several regions. As most sand dams have been constructed in the past five years, it is too early to draw firm conclusions about their impact. In particular, information on the benefit-costs and the demand for knowledge and governance for maintenance are lacking.

Floodwater harvesting systems

Some floodwater harvesting/spate irrigation systems in Ethiopia have been in use for several generations, but in almost all areas spate irrigation has developed in the last 30 years (van Steenbergen, 2011). According to recent estimates (reported by van Steenbergen, 2011) traditional spate irrigation in Ethiopia exceeds 100,000 ha (Alemehayu, 2008). Areas under improved or modernized spate irrigation stand at 20,000 ha and considerable investment is lined up: spate projects under design and construction exceed 50,000 ha (Alemehayu, 2008). Most systems are relatively small, with a few systems (Kobe, Yandefero, Dodota) touching the 4000 ha mark.

Examples of these water harvesting systems are known to exist in southern Ethiopia. The Ondokolo scheme (SNNPR) is reported (van Steenbergen, 2011) to have failed because of sedimentation and trash deposits. At this site, a gabion weir and gated offtake channel were constructed on a small steep sand bed river. The diversion was not well sited; that is, on a very sharp river bend, so virtually all the river flow was directed towards the canal intake. As a result the structure collected an enormous amount of flotsam and ceased to be operational. The Yandefero system (SNNPR) consists of a multitude (29) of short flood intakes which sustain a mixed cropping system of maize, sorghum and cotton. Farmers are mainly smallholders. They do not reside in the lowland area for fear of malaria and trypanosomiasis, preferring to live in the highlands. The entire area that can theoretically be irrigated is close to 4000 ha. Eleven of the flood intakes date back 30 years or more. Most of the remaining ones were developed in the last few years under a food-for-work (FFW) formula. Recently the Yanda River has started to degrade dramatically – going down one to two metres in large stretches. This has made it difficult to extend the flood channels and the majority of the intakes are not in use. Quantitative information on the benefit-costs and the demand for knowledge and governance for maintenance are not clear.

Integrated micro-catchment (MERET) approach

MERET is aimed at improving food security and livelihoods of chronically food insecure and impoverished communities, by providing food-for-work incentives to enable local people to invest in land management (and water harvesting) practices. Over the years, it has covered more than 600 sub-watersheds, each with 300 to 2000 participating households, in 74 *woredas* (districts) in six regions (including SNNPR and Oromia) and has rehabilitated over 400,000 ha of heavily degraded lands.

The origins of MERET can be traced back to 1980 when watershed rehabilitation was started as a famine response measure under a top-down large watershed approach. The switch to the current community-based micro-watershed approach occurred in 2003. The combination of soil and water conservation measures in both cultivated and uncultivated lands, gully rehabilitation, area closure and reforestation activities have resulted in significant reductions in soil loss rates and improvements in soil depth in the project sub-watersheds. Soil and stone bunds, *fanya juu* bunds, cut-off drains, check-dams, sediment storage dams, waterways and grass strips are some of the widely applied technologies. In cultivated fields erosion control measures are supplemented by soil fertility improvement measures such as use of legumes and compost so as to improve land productivity and generate immediate benefits to land users.

Current plans are that MERET should scale up to cover 'critical watersheds', as is envisaged in the current phase and scale out to additional communities. MERET is said to represent a model local level response to adapt to the adverse impacts of climate change, and it also has important climate change mitigation roles. Although some impact studies have been documented, the absence of reliable baseline information showing the nature and extent of problems before MERET is a constraint. A recent evaluation (Bewket, 2009) indicates that household level water harvesting technologies are valued by community members and provide an entry point for other interventions.

Conclusion

Ethiopia was not covered by the original World Bank commissioned SSWHS study (Reij et al., 1988; Critchley et al., 1992) but the experience of water harvesting mirrors that in countries which were studied. There is evidence of traditional practices that may be (i) implemented at household level using local runoff, and (ii) implemented at community level using runoff harvested from more extensive catchments. Over the last 40 years there have been many interventions by external agencies (both government and NGOs) to promote the spread of indigenous techniques or to introduce new water harvesting and soil and water conservation technologies. There are bright spots where these interventions have delivered positive impacts on livelihoods. There is also clear evidence that weaknesses identified by Kruger et al. (1996) have persisted. Poor performance can often be attributed to lack of appreciation of indigenous knowledge and practices by 'experts' and policy makers, and limited attention to maintenance and governance requirements in relation to local capacities after completion of the project. This is then compounded by failure to adopt a participatory approach in planning, designing and implementing interventions.

This chapter uses four techniques and approaches which have been implemented in the southern part of Ethiopia to illustrate the developments in these kinds of projects over recent decades. This evidence indicates that:

1   Many projects have been implemented in the past decade in Ethiopia; however, the effectiveness is not clear as impact assessments have not been carried out.

2   Multiple factors appear to influence success but in all cases adoption is mediated by external intervention by government agencies or NGOs.
3   Ponds are relatively simple to construct and maintain. Experience shows that an accelerated programme of pond building can be achieved with existing technical capacity. The economic feasibility of this technique and its long-term sustainability can be properly assessed only after further impact studies.
4   Sand dams have been delivering additional water in eastern Ethiopia since the 1970s, but it is only since an international consortium introduced them in the south that they have spread to other parts of Ethiopia. Factors playing a role here are the complexity of the technique and management of the water hampering spontaneous adoption and the need for an external promoter.
5   MERET provides an example of a long-term intervention which evolved from top down to community based approach during its 30 years of project life. A large number of micro-watershed sites provide examples of successful water harvesting techniques and integrated approaches. The question now is whether this experience can be replicated. Past success can be attributed to the intensive support by outside experts. Is it reasonable to expect other sites to sustain comparable activities with less intensive external help?
6   There is a long tradition of floodwater harvesting in some parts of Ethiopia but recent attempts by government and NGOs have been problematic. Impact assessments on the systems are mostly qualitative, and the complexity of the systems also demands technical support from government or donor organizations. A case can be made for better integration with the irrigation development strategy.
7   Compared to household techniques, catchment approaches have a wider impact and can improve water availability on a larger scale, including groundwater recharge. On the other hand, they have a higher demand for coordination and management. They are thus more expensive and complex, hampering spontaneous adoption, compared to household techniques.
8   Participation of local communities is important as the people have knowledge of local circumstances and they are the future users. Together with the technical experts, they should jointly decide on implementation, improving the functionality of the constructions and also the maintenance.

Note

1 This section has been largely compiled from FAO (2005).

References

Alemehayu, T. (2008) 'Spate profile of Ethiopia (A preliminary assessment),' Paper presented to FAO International Expert Consultant Workshop on Spate Irrigation, April 7–10, 2008, Cairo, Egypt.

Abay, F., Haile, M. and Waters-Bayer, A. (1999) 'Dynamics in IK: innovations in land husbandry in Ethiopia', *Indigenous Knowledge and Development Monitor*, vol. 7, no. 2, pp. 14–15.

Asrat, K., Idris, K. and Semegn, W. (1996) 'The flexibility of indigenous soil and water conservation techniques. A case study of the Hararghe area', in: C. Reij, I. Scoones and C. Toulmin (eds) *Sustaining the Soil, Indigenous Soil and Water Conservation in Africa*, Earthscan, London, UK.

Awulachew, S. B., Merrey, D. J., Kamara, A. B., Van Koppen, B., Penning de Vries, F., Boelee, E. and Makombe, G. (2005) *Experiences and Opportunities for Promoting Small-scale/Micro Irrigation and Rainwater Harvesting for Food Security in Ethiopia*, Colombo, Sri Lanka: IWMI (Working Paper 98).

Ayele, G., Ayana, G., Gedefe, K., Bekele, M., Hordofa, T. and Georgis, K. (2006) *Water harvesting practices and impacts on livelihood outcomes in Ethiopia*, Ethiopian Development Research Institute, Research Report VI, Addis Ababa, Ethiopia.

Bewket, W. (2009) *Rainwater harvesting as a livelihood strategy in the drought-prone areas of the Amhara region of Ethiopia*, OSSREA, Addis Ababa.

Critchley, W., Reij, C. and Seznec, A. (1992) *Water Harvesting for Plant Production. Volume II: Case Studies and Conclusions for Sub-Saharan Africa*. World Bank Technical Paper Number 157, Africa Technical Department, World Bank.

Critchley, W. (2009) 'Soil and water management techniques in rainfed agriculture – state of the art and prospects for the future.' Background note, CIS, Amsterdam.

Dale, D. D. (ed.) (2010) *Sustainable land management technologies and approaches in Ethiopia*, SLMP, Natural Resources Management Sector, MOARD, Addis Ababa, Ethiopia.

Dawa, B. (2006) 'Experiences of Adama area development program in water harvesting technology', Workshop Proceedings: Experience of Water Harvesting Technology in East Shewa and Arsi Zones, February 2006.

Dixon, A. (2001) 'Indigenous hydrological knowledge in southwestern Ethiopia', *Indigenous Knowledge and Development Monitor*, vol. 9, no. 3, pp. 3–5.

EPA (1997) *Environmental Policy for Ethiopia*. Environmental Protection Authority in collaboration with the Ministry of Economic Development and Cooperation, Addis Ababa, Ethiopia.

FAO (2005) *Irrigation in Africa in figures, AQUASTAT Survey – 2005*, FAO Water Report 29, Ethiopia, pp. 219–232.

FDRE (Federal Democratic Republic of Ethiopia) (2001) *Initial National Communication of Ethiopia to the United Nations Framework Convention on Climate Change (UNFCCC)*, Ministry of Water Resources, National Meteorological Services Agency, Addis Ababa, Ethiopia.

Fox, P., Rockstrom, J. and Barron, J. (2005) 'Risk analysis and economic viability of water harvesting for supplemental irrigation in semi-arid Burkina Faso and Kenya', *Agricultural systems*, vol. 83, pp. 231–250.

Kebede, T. (1995) 'Experience in soil and water conservation in Ethiopia', Paper presented at Agri-Service Ethiopia's Annual Technical Meeting, Wondo-Genet, Ethiopia.

Kruger, H., Fantaw, B., Michael, J. and Kajela, K. (1996) 'Creating an inventory of indigenous soil and water conservation measures in Ethiopia', in: C. Reij, I. Scoones and C. Toulmin (eds), *Sustaining the soil, indigenous soil and water conservation in Africa*, Earthscan, London, UK.

Lasage, R., Aerts, J. C. J. H., Mutiso, G-C. M. and Vries, A. de (2008) 'Potential for community based adaptations to droughts: Sand dams in Kitui, Kenya', *Physics & Chemistry of the Earth*, vol. 33, pp. 67–73.

Michael, Y. G. and Herweg, K. (2000) 'From indigenous knowledge to participatory technology development', Centre for Development and Environment, University of Berne, Switzerland, 52 pp.

Mintesinot, B., Kifle, W. and Leulseged, T. (2005) *Fighting famine and poverty through water harvesting in Northern Ethiopia*. Comprehensive Assessment Bright Spots Project Case Study Report, available at: http://www.iwmi.cgiar.org/brightspots/index. asp, accessed 30 May 2012.

Reij, C., Mulder, P. and Begemann, L. (1988) *Water Harvesting for Plant Production*, World Bank Technical Paper No. 91, World Bank, Washington D.C.

Shiferaw, B. and Holden, T. H. (2001) 'Farm-level benefits to investments for mitigating land degradation: empirical evidence from Ethiopia', *Environment and Development Economics*, vol. 6, pp. 335–358.

Tesfai, M. and Graaff, J. de (2000) 'Participatory rural appraisal of spate irrigation systems in eastern Eritrea', *Agriculture and Human Values*, vol. 17, no. 4, pp. 359–370.

UNESCO (2011) 'Konso Cultural Landscape', UNESCO World Heritage List, available at: http://whc.unesco.org/en/list/1333, accessed 30 May 2012.

van Steenbergen, F., Lawrence, P., Salman, M. and Faurès, J. M. (2010) 'Guidelines on spate irrigation', FAO irrigation and drainage paper 65, FAO, Rome, Italy.

van Steenbergen, F., Haile, A. M., Alemehayu, T., Alamirew, T. and Geleta, Y. (2011) 'Status and potential of spate irrigation in Ethiopia', *Water Resource Management*, vol. 25, issue 7, pp. 1899–1913.

Watson, E. E. (2009) *Living Terraces in Ethiopia: Konso Landscape, Culture and Development*, James Currey (an imprint of Boydell and Brewer) East Africa Series, Woodbridge and New York.

Chapter 6

Kenya

From drought relief to business model

Alex Oduor, Kipruto Cherogony, Maimbo Malesu and William Critchley

Introduction

Kenya was selected as a key country for documentation under the Sub-Saharan Water Harvesting Study (SSWHS) because of its rich experiences of water harvesting (Critchley et al., 1992). Like Burkina Faso in the West African Sahel, Kenya hosted a number of water harvesting projects which sprang up in the 1980s as a response to droughts. There was the same enthusiasm for this 'new technology'. While Kenya has relatively little traditional water harvesting, the country has abundant possibilities for its practice. We look at the history of water harvesting in Turkana and Baringo Districts of the Rift Valley Province, where several initiatives were underway in the 1980s: several of these were examined under the SSWHS (Critchley et al., 1992). This chapter reviews the current situation. But the picture would not be complete without also reporting one of the most interesting recent developments: the evolution of road runoff harvesting in Eastern Kenya. A notable success that also emerged later in the country has been the sand dam programme; however this technology is discussed elsewhere in the book (see Chapter 5: Ethiopia).

Kenya: background

Kenya is often thought of as being a green and well-watered land: but this is only true for a minority of the country – within the highland zone. Rainfall is bi-modal for most parts with the 'long rains' commencing in March and ending in June, while the (usually more reliable) 'short rains' start in October and proceed to December. Annual average precipitation varies from 200 mm and below in northern and north-eastern regions of the country, which occupy approximately 40 per cent of total area, to south-easterly and upper portions of central rift valley regions whose precipitation ranges from 200 to 750 mm. These regions occupy approximately 30 per cent of Kenya. The remaining areas are the coastal and Lake Victoria belts and peripheries of the mountain ranges which occupy about 20 per cent of Kenya: these experience an annual average precipitation of approximately 750 to 1250 mm. Finally, the Aberdare and Mount Kenya regions occupy the remaining 10 per cent of Kenya, where rainfall surpasses 1250 mm per annum.

About 8 per cent of Kenya's landmass is arable with permanent crops occupying only 1 percent of the land. The rest of Kenya comprises rangelands or forest/bush areas. Irrigation is especially pertinent in the face of recurrent droughts, floods and prolonged dry spells, which cause food insecurity and famines in the country. But while Kenya has an irrigation potential of 539,000 hectares (based on surface water availability) only 110,000 ha have been developed: this represents less than 2 per cent of the total cropped area – around half of the average for Sub-Saharan Africa as a whole.

In the last population census carried out in 2009, the total number of people in the country was estimated at 38.6 million (KNBS, 2009). Furthermore Kenya's population growth rate remains one of the highest in Sub-Saharan Africa. The majority live in the relatively well-watered highlands, but a significant proportion – around a quarter – dwell in the 80 per cent of the land that is arid or semi-arid, where the rainfall to evapotranspiration ratio falls below 0.6. In these areas, rainfed farming is fraught with risk. Kenya lies within the Great Horn of Africa region which is prone to cyclic drought. According to Mateche (2011), the drought cycle in Kenya has become shorter, with droughts becoming more frequent and intense due to global climate change and environmental degradation. The cycle has shortened from every ten years, down to every five years, further to every two to three years, and currently each year is characterized by some dry spells. This is one reason why water harvesting is so crucial for Kenya.

History and importance of water harvesting practices

Turkana

Kenya's potential for water harvesting had been noted as long ago as the 1950s, when trials were set up in Turkana (Cullis, 1981; Finkel, 1986; Erukudi, 1991). Though said to have failed, there was sufficient promise to inspire a number of new experimental schemes under the Range Management Division in the 1960s and 1970s. The 1961 drought and resultant famine had brought Turkana to national attention. A number of small-scale irrigation schemes were established, but with little success – being expensive to establish and difficult to manage and maintain. Water harvesting was the alternative. The best documented of these was the 40 ha 'Impala Pilot Water Spreading Scheme', established at Lorengippe in southern Turkana. This was heralded in the subtitle of the project report as: 'A hope for an impoverished people' (Fallon, 1963). However as Finkel (1986) reports, the project that was intended to produce both sorghum and fodder failed, not so much for technical, but more for sociological reasons.

During the 1980s, water harvesting returned to Turkana from two different sources. According to Critchley (1999), who together with Erukudi reviewed the legacy of Turkana's water harvesting experience in the 1980s:[1]

> On the one hand it was developed by small NGOs and Church Missions as a chosen activity, developed in discrete areas with food-for-work often used

for support. Foremost in this field was the Turkana Water Harvesting Project. On the other hand it was implemented, large-scale, by the Turkana Rehabilitation Programme, where the driving force was the need to maintain distribution of food-for-work, utilising rainwater harvesting as one of the main vehicles.

The Turkana Rehabilitation Programme (TRP) was established in 1980 with the mandate of getting the ongoing famine under control. There was a massive food distribution programme which reached about 80,000 people (40 per cent of the district's population). But then a fundamental question arose: what was the best activity to fulfil the 'work' element of food-for-work? TRP answered this by picking up the challenge of water harvesting on a huge scale, supported by up to three-quarters of its food distribution at one time. Logistics dictated that it was easiest to cluster people – men and women – and engage them in earth moving activities. All work was done by hand: digging, carrying and compacting earth bunds (see Photo 6.1). Blocks of V-shaped 'negarim' tree microcatchments provided a simple one-person/one-day/one-ration formula, but with no maintenance and no protection, tree survival rates were dismal. According to Finkel (1986) the experience with water harvesting bunding was no better: between 1983 and 1984, 120 km of earth bunds were established using 20,000 labourers rewarded with food-for-work. But there was no proper surveying carried out and these bunds served 'no useful purpose other than making work' (Critchley, 1999).

The first technical input into the programme was under the auspices of FAO in 1985. The 'trapezoidal bund' was the keystone of the resultant design manual

Photo 6.1 Women compacting trapezoidal bund with wooden poles, Turkana: 1980s
(W. Critchley)

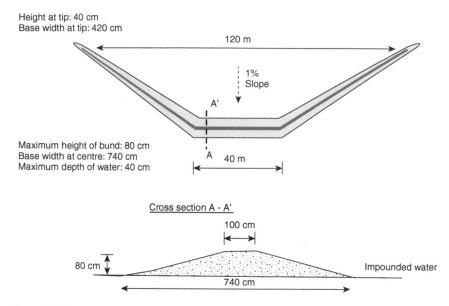

Height at tip: 40 cm
Base width at tip: 420 cm

120 m

1%
Slope

A'

Maximum height of bund: 80 cm
Base width at centre: 740 cm
Maximum depth of water: 40 cm

A   40 m

Cross section A - A'

100 cm

80 cm

Impounded water

740 cm

Figure 6.1 Trapezoidal bund (adapted from Finkel et al., 1986; after Critchley, 1999)

(See Figure 6.1) (Finkel et al., 1987). This is an external catchment system that works through capturing overland flow. It is similar in many ways to the traditional *teras* of Sudan's Border Region (see Chapter 9). It is not recorded how many of these structures were built; this is unsurprising as the TRP had a weak monitoring system that only measured the basics of food distributed and cubic metres of earth moved. But things changed quickly with the appointment of a District Rainwater Harvesting Officer. This signalled a move in direction away from trapezoidal bunds and towards spate irrigation (or floodwater harvesting). Of four schemes constructed, one – at Karubangarok – represented the nadir of the entire water harvesting programme under TRP. Enormous bunds were constructed to divert a *lugga* (an ephemeral water course), but with no credible technical design. The first floods tore through the scheme, breaching bunds and creating deep gullies. As Critchley (1999) wrote after visiting the site some years later: 'The scheme was abandoned and the skeleton of the bunds remains today as a monument to mismanagement of food-for-work'.

The Turkana Water Harvesting Project (TWHP)[2] emerged in 1985 from an earlier water harvesting project established in 1979. The manager of that original initiative, the Lokitaung Water Harvesting Project visited rainwater harvesting sites in Israel with local counterparts. Early technical results were promising. The project's main objectives were to 'improve and demonstrate appropriate rainwater harvesting techniques for crop production while developing better systems of food-for-work'. TWHP modified the trapezoidal bund originally designed for use under TRP, and simultaneously reduced food-for-work rations to a point that these were supplied as a capital grant constituting 15 to 20 per cent of a 'full' ration. While operating on a much smaller scale than TRP, and without the same type of pressures, by 1990 TWHP had succeeded in stimulating the construction of 200 water harvesting plots of which the large majority were utilized and maintained on a family basis (Critchley, 1999).

Baringo

Baringo District was likewise a 'laboratory' for water harvesting in the 1980s, though this was not driven by the need to provide food-for-work. Three main initiatives began at that time and provided case studies for the SSWHS (Critchley et al., 1992). The Baringo Pilot Semi-Arid Area Project (BPSAAP) was an initiative supported by the World Bank and implemented through the Government of Kenya's Ministry of Agriculture and Livestock Development. This project's mandate was broad but it aimed specifically to 'establish a field-tested basis for the rehabilitation and development of the semi-arid areas of the Baringo District' focusing on soil and water conservation, water harvesting and land rehabilitation (MoALD, 1984).

Under this project, various systems of water harvesting were tested, including hand-dug contour ridges specifically for crops (see Photo 1.1 in Chapter 1; see also Figure 2.2 in Chapter 2), mechanized contour ridges (rather larger structures; made by tractor with a disc plough) and semi-circular bunds (similar to the *demi-lunes* reported in Chapter 4 on Burkina Faso and Chapter 7 on Niger). The latter two systems were developed for rangeland rehabilitation with trees and grass. At the time of the project trials, the technical performances were positive (in some cases excellent, given the high runoff coefficients, the rich soils, and consequent vigorous plant response to extra water), but the key unanswered challenges were poor management of rehabilitated demonstration sites post-project, and lack of voluntary adoption of technologies by the agropastoralists who lived in the area (MoALD, 1984; Critchley et al., 1992).

Two other projects, namely the Baringo Fuel and Fodder Project (BFFP), and the Fuelwood Afforestation Extension Programme (FAEP) were featured in the SSWHS and their water harvesting initiatives are reported in Critchley et al. (1992). As noted in this report, the two projects used basically similar techniques for tree planting: contour bunds (with upslope furrows) spaced some 5 to 10 m apart, constructed by motor grader or tractor. Trees were planted in/close to the furrows where the overland flow collected. As their names suggest, the projects' emphases were firmly on trees, and, to a lesser extent, grass and fodder. It was further noted under the SSWHS that neither initiative had moved out of the controlled project-managed phase. Doubts were also cast on the long-term viability of mechanized techniques in the area and the wisdom of introducing exotic *Prosopis* spp. In this respect the report considered that this tree would 'almost certainly become naturalized in the area, with mixed (though on balance positive) benefits judging from experience elsewhere in Africa' (Critchley et al., 1992).

With the benefit of hindsight, the word 'positive' should almost certainly now be replaced by 'negative'. A recent and critical review of the introduction of this tree has once again demonstrated how the best intentions can prove to have unforeseen and unfortunate consequences (Mwangi and Swallow, 2008).

Key developments

Institutional

Since the earliest days of water harvesting initiatives in Kenya, there have been various important institutional developments. While there was already a Soil and Water Conservation branch within the Ministry of Agriculture, a specialized water harvesting unit was set up within that branch during the 1990s when the 'catchment approach' to soil and water conservation was introduced. That marked an acknowledgment of water harvesting's 'legitimacy'. The Regional Soil Conservation Unit (RSCU) was established in 1982 as a special project under Sida (Swedish International Development Authority: formerly abbreviated to SIDA). When the RSCU moved out of the Swedish Embassy and became RELMA (the Regional Land Management Unit hosted at the International Centre for Research in Agroforestry, ICRAF)[3] in 1998, it took on a stronger role in the field of promoting water harvesting – through training but especially in terms of publishing and dissemination of materials. This gave Kenya's water harvesting efforts a stronger knowledge base and a further boost.

Currently there are myriad players involved in the promotion of water harvesting best practices in Kenya's ASAL districts. The key players can be categorized into the following groups: governmental agencies, international organizations, non-governmental organizations (including religious groups) financial institutions, networking agencies and local community support agencies. The Kenya Rainwater Association (KRA) and the umbrella organization the Southern and Eastern Africa Rainwater Network (SearNet, hosted, like RELMA before it, at ICRAF), are the two central organizations playing a capacity and networking role for water harvesting in Kenya. SearNet, and its national representative, KRA, have both spearheaded water harvesting since 1998. KRA has grown into an autonomous body that is able to solicit funds to support field activities. Together, SearNet and KRA publish regular information via electronic and print media to practitioners across the country – and in SearNet's case, throughout the region.

Trapezoidal bunds: Turkana and elsewhere

In Kakuma, within Turkana, trapezoidal bunds (TBs) continue to be the accepted design, and the Ministry of Agriculture's National Agriculture and Livestock Extension Progamme (NALEP) selected a 500 ha site to showcase the technique, and there will soon be other examples of such initiatives in the ASAL regions, to be financed by the Swedish Development Cooperation Agency. Design and layout is carried out by the District FFA Coordinator with support from various

Table 6.1 Original design dimensions for trapezoidal bunds in Turkana district

Slope percentage	Length of base bund (m)	Length of wing wall (m)	Distance between tips (m)	Earth work per bund (m³)	Cultivated area per bund (m²)
0.5%	40	114	200	355	9600
1.0%	40	57	120	220	3200
1.5%	40	38	94	175	1800

Source: Critchley and Siegert, 1991; derived from Finkel et al., 1987

other agency staff where necessary. The World Food Programme finances the initiative through the contribution of food as an incentive for the farmers to engage in construction. Farmers' contributions include fencing, maintenance and general management of the structures. Their contribution is in kind. According to the National Food-for-Assets Coordinator, Mr Paul Kimeu (personal communication), trapezoidal bunds have been established in Turkana North, Turkana South, Turkana West and Turkana Central Districts. All the TBs in Turkana are established on trust land.

A visit by SearNet/ICRAF scientists in 2011 verified that the design of the bunds are still determined by the guidelines developed in 1987. These dimensions are given in Table 6.1. for groundslopes of 0.5, 1.0 and 1.5 per cent.

The food-for-assets (FFA) initiative of the World Food Programme (sometimes still known by its former name; food-for-work) has brought together important stakeholders from government and external support agencies alike with the main objective of protecting and rebuilding livelihoods in the arid and semi-arid lands. At the national level, it is coordinated with the Office of the President's Special Programmes, the Ministry of State for the Development of Northern Kenya and other Arid Lands and/or the newly established National Drought Management Authority, and the Ministries of Agriculture, Water and Irrigation, Environment and Mineral Resources, and Livestock Development (WFP, 2012). The six key outcomes of the FFA demonstrate how it has moved on from simply emergency relief towards building resilience to cope with drought:

- to increase pasture and browse conditions;
- to improve diversification of food sources through on-farm projects, greenhouses, *zaï* pit technology and other microcatchments, and planting of fruit trees;
- to improve access of water for domestic and livestock use;
- to reduce environmental degradation through check dams, terraces, and so on;
- to improve access to markets, schools, and so on, through construction of feeder roads; and
- to improve capacity of community to implement food security projects through training, technology transfer and field days or exposure visits.

It is noteworthy that the second objective is directly related to water harvesting for crop production and that it also takes cognisance of a technology from West Africa – *zaï* planting pits (see Chapter 4: Burkina Faso). Nevertheless, the emphasis is still on provision of full subsidies for construction.

During Phase One of the Protracted Relief and Rehabilitation Operations (PRRO) for the period May 2009 to April 2012, the programme covered various semi-arid and arid (ASAL) areas, including Isiolo, Kajiado, Turkana, Marsabit, Mandera, Garissa, Wajir, Taita Taveta, Kitui, Kwale, Makueni, Tana River, Moyale and Samburu Districts. In Makueni specifically, the FFA programme installed 34 trapezoidal bunds of which 14 were located in Kibwezi and 20 in Kathonzweni divisions. According to Mr Joel Mutiso (personal communication), the Kibwezi District Food for Assets Coordinator, their performance has encouraged farmers who are still new to this technology. All these structures are established on individually owned lands, although not everybody has title deeds as yet. Farmers 'contribute' by providing labour during construction which is compensated through provision of food as an incentive. Phase Two of the PRRO programme commenced in May 2012 and will continue till April 2015 covering 16 ASAL districts.

Eastern Kenya

Tapping runoff from road drainage is one of the most ingenious forms of water harvesting, yet the Sub-Saharan Water Harvesting Study was unaware of this phenomenon in the 1980s. That is unsurprising as the practice was relatively young, still under evolution, and unreported. In the years after that study, some attention was raised to this local innovation (e.g. Mutunga, 2001, 2002; Mutunga and Critchley, 2001; Ngigi, 2003; Nissen-Petersen, 2006). In principle, road runoff harvesting makes use of road drainage runoff either from the crest of the road into a mitre drain, or from the far side of the road through a culvert. The former provides less water (usually) though is simpler to control. As with all forms of water harvesting, by harnessing runoff not only is it productively used, erosion is prevented or at least reduced. Road construction is notorious for causing gullies through farmland by irresponsible drainage design, and that phenomenon is not unique to Kenya. Rwanda has particularly excelled in incorporating silt traps along the mitre drains for efficient road runoff conveyance.

Farmers in Eastern Kenya – and especially within Machakos, Kitui and Mwingi Districts – are pioneers in terms of road runoff harvesting (RRH). One can only speculate why this might be, but farmers in this area have well-developed terracing systems (using the *fanya juu* design where the furrow is downslope of the bund) and thus runoff can be readily channelled into the furrows and circulated around the farm. The furrows then act as infiltration ditches. Mutunga and Critchley (2001) documented the case of Mr Musyoka Muindu from Mwingi, who has skilfully devised a system that leads water through his farm from a road culvert – and also from a hillside above his farm. The SASOL Foundation Kenya plans to upscale such best practices in the adjacent Kitui District.

In Uganda the utilization of RRH to provide extra water to bananas is common also; one such example is pictured on the cover of this book. A particularly creative way of using harvested water is to use it to carry cattle manure to fruit trees: this farmer-designed innovative 'fertigation' system from Uganda is also described amongst case studies in Mutunga and Critchley (2001). Box 6.1 gives more information about Uganda and water harvesting.

A survey in 2011 by scientists from SearNet/ICRAF uncovered several examples of the further development of RRH in Eastern Kenya. One particular example from Machakos District is summarized here. Mr Samuel Maingi, a 64-year-old former chief, has a road runoff harvesting system that he devised for himself – with some technical support from the Ministry of Agriculture (see Photo 6.2). Previously, road runoff adjacent to his plot found its way directly to a stream below his farm, causing damage to the unpaved road on its way.

Box 6.1 Water harvesting for banana production in Uganda

Despite the relatively high rainfall regimes in Uganda, water infiltration trenches/ditches, which catch and hold water from paths, roads or natural waterways, have become an important part of the banana production system in the last two decades. This is partially explained by production of *matooke* cooking bananas, the staple food in Uganda, moving into ever more marginal areas. Uganda is a country endowed with an equatorial climate. However rainfall varies from 750 mm per year in the north-east Karamajong pastoral areas to 1500 mm per year in the humid zones. The drylands of Uganda, the 'cattle corridor', are prone to drought: yet they are producing more and more bananas. In total 28 per cent of Uganda's farmland is taken up by bananas, overshadowing cereals at 26 per cent and root crops at 17 per cent. Based on farmer-designed innovative systems, water infiltration trenches/ditches, dug along the contour, at specified intervals, for retaining runoff in banana plantations were promoted by the Uganda Soil Conservation and Agroforesty Pilot Project (USCAPP) from 1992 to tackle this problem (Rockström, 2000; Kiggundu, 2002). Water harvesting, especially for banana production, was also part of 'conservation agriculture' introduced into Uganda through two pilot projects, a FAO Technical Cooperation Project (TCP/UGA/2903) in eastern Uganda (Pallisa and Mbale Districts) in 2002 and a Sida-funded project in western Uganda (Mbarara District) in 2000 (Nyende et al., 2007). In combination with other complementary management practices such as weeding, adding manure, mulching, correct spacing and variety choice, as well as pest and disease control, water harvesting has certainly increased the production of bananas bringing more secure food supplies and improved rural incomes. This is the 'water harvesting plus' (WH+) introduced in Chapter 2.

Photo 6.2 Pumping water from Mr Samuel Maingi's road runoff harvesting pond
(W. Critchley)

Mr Maingi determined to capture this water for his own agricultural use, and
diverted it into a pond that he constructed; 5 × 10 m wide and 4 m deep, lined
with 0.8-mm-thick high-quality PVC that costs US$4.7 m² in Nairobi. The pond
was constructed over a period of one year, in 2004, with a ten-year guarantee and
wasn't leaking some six years later. The layout is sketched in Figure 6.2.

According to Mr Maingi, 'one good rain shower' is enough to fill the whole
pond. He pumps the water with a hand operated rope-and-washer pump (see
Figure 6.2) up to four terrace levels or 3 m higher and applies supplementary
irrigation through a hosepipe to part of his 1.5 ha plot with the ponded water.
He grows, and waters, fruits and vegetables including citrus trees, pawpaw,
mangos, *sukuma wiki* (kale: used as a local vegetable), cabbage, onions, tomatoes,
sweet potatoes, peppers, eggplants, cowpeas, and chickpeas. He uses fertilizer
and manure to improve his relatively infertile land.

Mr Peter Ndonye Wambua, Maingi's neighbour and a member of the Mango
Integrated Community Based Organization, a co-operation of local smallholder
farmers, adopted this system of RRH using a lined pond, but in addition linked
this to a greenhouse that is managed by this group of farmers (see Figure 6.3).
The greenhouse is a tripartite arrangement between World Vision International
who coordinate the promotion of this system, Amiran Kenya Ltd (from Israel), a
private company that gives technical support in irrigation infrastructure and
Equity Bank which provides finance. Maintenance of the ponds is carried out by
farmers who were trained in 2004 by Kenya's Regional Land Management Unit

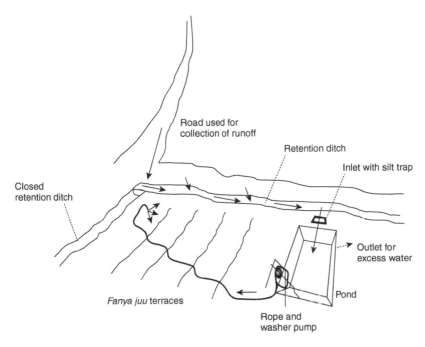

Road used for
collection of runoff

Retention ditch

Inlet with silt trap

Closed
retention ditch

Outlet for
excess water

Fanya juu terraces

Pond

Rope and
washer pump

Figure 6.2 Schematic overview of the road runoff harvesting on Samuel Maingi's farm

(RELMA). Greenhouses are a new concept in the district. This arrangement comprises a 'package': a rope-and-washer pump, greenhouses for drip irrigation, a collapsible synthetic water tank, and a plastic lined pond of 250 m^3 capacity (see Figure 6.3). Farmers excavate the pond; they manage the greenhouse and horticultural crops therein – usually tomatoes or sweet peppers.

Currently 15 farmers have adopted a greenhouse, 11 of them with the support of World Vision International (WVI; coordination), Kenya Equity Bank (funding) and Amiran Ltd (technical input); four have done so independently. While currently there are subsidies available, the eventual aim is to make this a self-standing private commercial exercise. For this to become a true 'business model' (see Chapter 3), it is necessary to first develop a standard business plan to estimate the return on investment, through a cost-benefit analysis using various indicators based on different scenarios (amount of water available, number of ponds per farmer or community, expected revenue from greenhouse crops, etc.).

A complete greenhouse package (infrastructure and training during the first season) costs US$1750–2350. Tomatoes are the main crop produced currently in the greenhouses; demand is high, and prices good – this area is situated close to a tarmac road and a short distance from the busy market town of Machakos. To facilitate further adoption, tailor-made options must be readily available for farmers.

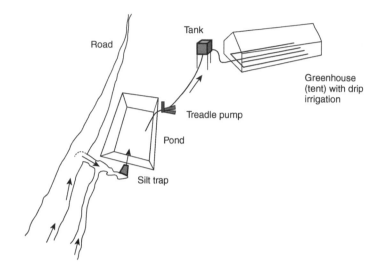

Figure 6.3 Schematic overview of the road runoff harvesting–greenhouse system

With regard to the upscaling of the use of ponds, lessons can be learned from Rwanda; ironically this has occurred after the Rwandans themselves learnt the basics from a visit to Kenya organized through SearNet in October 2007. The Rwandese copied the concept and began to implement it on a large scale. This has been made possible by the direct support provided by the government.

Concluding discussion

From being an exciting, but experimental technology in the 1970s and 1980s, associated with drought relief, Kenya's water harvesting profile and portfolio has matured. Initially the proponents of water harvesting had to struggle to make their voice heard: they were even occasionally derided. And the failures of the food-for-work schemes under the Turkana Rehabilitation Programme did not help the cause. But TRP's mistakes have to be seen in perspective. This occurred because poor planning failed to link sensible 'work' with the food that had to be distributed. The Turkana Water Harvesting Project managed to develop a much more viable system, both in terms of food and technology. There have also been false dawns: under the Baringo Pilot Semi-Arid Area Project, the technologies that delivered excellent technical results failed to appeal to the local land users. And other projects in that area unwittingly introduced an invasive species, *Prosopis juliflora* just as projects in Sudan and Somalia were doing at the same time. But what do we see now in Kenya? There is a strong programme of trapezoidal bunding being currently promoted, though once again regular subsidies for construction have become the order of the day. Time will tell if these structures will be maintained by farmers. However the most interesting development to many will be the development of the 'business model' which connects road runoff harvesting with greenhouses and the market. It may not be applicable everywhere, but it certainly will have a sustainable future close to hungry urban centres.

Notes

1 The original review was commissioned by the Intermediate Technology Development Group (ITDG) and carried out by Critchley and Erukudi in 1992. An internal report was used with permission as the basis for Critchley (1999).
2 The activities of TWHP were filmed in 1990 under an initiative that led to 'Looking After Our Land': a booklet and film module (Critchley, 1991). The film can be viewed on http://www.thewaterchannel.tv/en/videos/categories/viewvideo/866/agriculture/looking-after-our-land (the same film features construction sequences of water harvesting technologies from Burkina Faso, namely stone lines, zaï and permeable rock dams).
3 ICRAF has now become the 'World Agroforestry Centre'.

References

Critchley, W. (1991) *Looking After Our Land* (Booklet and Film: also in French as *Pour Proteger Nos Terres*), Oxfam Publications, Oxford.
Critchley, W. and Siegert, K. (1991) *Water Harvesting*, Food and Agriculture Organisation, Rome.
Critchley, W. (1999) 'Food-for-work and rainwater harvesting: Experience from Turkana District, Kenya in the 1980s', in: D. W. Sanders, P. C. Huszar, S. Sombatpanit and T. Enters (eds), *Incentives in Soil Conservation: from theory to practice*, World Association of Soil and Water Conservation, Oxford and IBH Publishing, New Delhi.
Critchley, W. and Erukudi, C. (1992) *Review of water harvesting activities in Lokitaung Division, Turkana District, Kenya*, Internal Report, Intermediate Technology Development Group, Rugby, UK.
Critchley, W., Reij, C. and Seznec, A. (1992) 'Water harvesting for plant production. Volume 2. Case Studies and Conclusions for Sub-Saharan Africa', World Bank Technical Paper no. 157, Washington D.C.
Cullis, A. (1981) 'Water harvesting philosophy related to run-off farming', Mimeo, Intermediate Technology Development Group (ITDG), Rugby, UK.
Erukudi, C. E. (1991) 'Investigations into small-scale irrigation and water harvesting management strategies for the dryland Turkana District, Kenya', MSc Thesis, Silsoe College, Silsoe, UK.
Fallon, L. E. (1963) 'Water spreading in Turkana: A hope for an impoverished people', Mimeo, USAID, Nairobi.
Finkel, M. (1986) 'Water harvesting in Turkana, Kenya', Paper presented to the World Bank workshop on water harvesting, Baringo, Kenya, November 1986. Summarized in Critchley, W. (1986) *Some Lessons From Water Harvesting in Sub-Saharan Africa*, Africa Technical Division, World Bank, Washington D.C.
Finkel, M., Erukudi, C. and Barrow, E. (1987) *Turkana Water Harvesting Manual*, Mimeo, Ministry of Agriculture, Nairobi.
Kiggundu, N. (2002) *Evaluation of rainwater harvesting systems in Rakai and Mbarara Districts, Uganda*, GHARP case study report, Greater Horn of Africa Rainwater Partnership (GHARP), Kenya Rainwater Association, Nairobi.

KNBS (Kenya National Bureau of Statistics) (2009) *Kenya 2009 Population and Housing Census*, Kenya National Bureau of Statistics, Nairobi.

Mateche, D. E. (2011) 'The Cycle of Drought in Kenya: a Looming Humanitarian Crisis', Environmental Security Programme, Institute for Security Studies, Nairobi, available at: http://www.iss.co.za/iss_today.php?ID=1217, accessed 31 May 2012.

MoALD (Ministry of Agriculture and Livestock Development) (1984) *Baringo Pilot Semi-Arid Area Project (BPSAAP) Interim Report*, Government of Kenya, Nairobi.

Mutunga, K. (2002) 'Road Runoff Harvesting', Technical leaflet, RELMA, Nairobi.

Mutunga, K. (2001) 'Water conservation, harvesting and management (WCHM) – Kenyan experience', in: D. E. Stott, R. H. Mohtar and G. C. Steinhardt (eds), Sustaining the global farm, selected papers from the 10th International Soil Conservation Organisation Meeting, May 24–29, 1999 at Purdue University and the USDA-ARS National Soil Erosion Research Laboratory.

Mutunga, K. and Critchley, W (2001) *Farmers' Initiatives in Land Husbandry: Promising Technologies for the Drier Areas of East Africa*, RELMA Technical Report No. 27, Nairobi.

Mwangi, E. and Swallow, B. (2008) '*Prosopis juliflora* invasion and rural livelihoods in the Lake Baringo Area of Kenya', *Conservation Society*, vol. 6, no. 2, pp. 130–140.

Ngigi, S. N. (2003) *Rainwater Harvesting for Improved Food Security: Promising Technologies in the Greater Horn of Africa*, Kenya Rainwater Association, Nairobi.

Nissen-Petersen, E. (2006) 'Water from roads: a handbook for technicians and farmers on harvesting rainwater from roads', Danida, Kenya.

Nyende, P., Nyakuni, A., Opio, J. P. and Odogola, W. (2007) 'Conservation agriculture: a Uganda case study', Nairobi, African Conservation Tillage Network, Centre de Coopération Internationale de Recherche Agronomique pour le Développement, Food and Agriculture Organisation, Rome.

Rockström, J. (2000) 'Water Resources Management in Smallholder Farms in Eastern and Southern Africa: An Overview', *Physics and Chemistry of the Earth, Part B: Hydrology, Oceans and Athmosphere*, vol. 25, no. 3, pp. 275–283.

WFP (World Food Programme) (2012) 'Protracted Relief and Recovery Operations – Kenya 200294. Protecting and Rebuilding Livelihoods in Arid and Semi-Arid Areas', Executive Board First Regular Session Rome, 13–15 February 2012, projects for executive board approval, Agenda item 8.

Chapter 7

Niger

Small-scale and simple for sustainability

Sabina Di Prima, Abdou Hassane and Chris Reij

Introduction

In Niger, as in other Sahelian countries, food security still remains a top priority. Since the 1970s, Niger has suffered a series of food crises which have resulted from several factors, the most important being an annual increase in population (3.3 per cent) greater than agricultural growth (estimated at 2.5 per cent) and the trend towards a drier climate (Ministère de l'Eau, de l'Environnement et de la Lutte Contre la Désertification, 2011). The combination of these factors has inevitably led to increased pressure on the environment, a disruption of ecological balance and land degradation. In many cases, land has been exploited beyond its capacity with a consequent loss of its productive potential.

To counteract these negative trends, in the 1980s, several donor agencies in collaboration with government counterpart agencies began water harvesting projects to restore degraded land to productivity. These projects were largely concentrated in Niger's Tahoua Department, which was characterized by the topographical pattern of fertile valleys divided by barren, degraded plateaus which produced considerable runoff. The type of techniques promoted and the approaches used differed considerably. From the middle of the 1980s, heavy machinery was used, in combination with village labour mobilized with food-for-work (FFW), to construct water-harvesting bunds on the plateaus. Hand-dug trenches were also introduced on the slopes of these plateaus. In contrast, towards the end of the 1980s a soil and water conservation project, funded by the International Fund for Agricultural Development (IFAD), began promoting simple water harvesting techniques such as semi-circular bunds (*demi-lunes* in French) and planting pits (termed *tassa* in Hausa). Through a focus on these four technologies, this chapter reflects on the main developments in the field of water harvesting in Niger over the last three decades. Particular emphasis is given to *demi-lunes* and *tassa*, because they continue to be used, adapted and expanded by farmers without external support.

Niger: background[1]

Niger is a vast plateau situated at an average altitude of 500 m extending over 1267 million km². The northern part of the country is covered by desert and arid mountains, and only the southern fringe of the country has enough rainfall to allow reliable agriculture (IFAD, 2009). It has a predominant Sahelian climate characterized by a long dry season (October to May), a short rainy season (June to September) and a significant variation in the annual rainfall from north to south (from less than 100 mm to 700–800 mm). About two-thirds of the territory is located in the Saharan zone (rainfall below 100 mm/year). The remaining part of the country is divided between the Sahelo-Saharan zone (annual rainfall between 100 and 300 mm) and the Sahelian zone (where annual rainfall ranges from 300 to 600 mm). Only a small fraction of the south-western part of the country receives more than 600 mm rainfall. Rainfall is characterized by high variability, both in time and in space. It is erratic and often occurs in isolated showers. The average annual temperature reaches 29°C. The potential annual evapotranspiration is very high (2114 mm/year in the area of Tillabéry) and exceeds precipitation with the exception of the month of August (FAO, 2005).

The lifeline of the country is the Niger river, which flows year-round across the south-western part of the country for about 560 km from north-west to south-west. After two decades of deliberations and negotiations with donors and other riverine countries, Niger began the construction of the Kandadji dam in 2011; this will expand the area under irrigation in the next two decades by an estimated 30,000 ha. Furthermore, thanks to this dam, the country's dependence on electricity imported from Nigeria will decrease.

In 2010, the population of the country was approximately 15.5 million, of whom about 83 per cent lived in the rural areas (FAOSTAT, 2010). The average population density is about 10 inhabitants per km². However, 90 per cent of the population is concentrated in the southern strip (about 200 km wide) along the border of Nigeria where rainfall is favourable both to agriculture and to agropastoralism. Niger is one of the world's poorest countries with a GDP per capita of US$217 in 2003. Approximately two-thirds of the population lives below the poverty line, and one-third below the extreme poverty line (IFAD, 2007). Women and pastoralists are particularly vulnerable, although the 2005 crisis revealed high poverty levels in dense populated agricultural areas, in particular among farmers with very small landholdings and among landless households (IFAD, 2007).

The agricultural sector continues to occupy a central place in Niger's economy. It employs 87 per cent of the workforce and contributes 40 per cent of the GDP (FAO, 2005). Rainfed agriculture is dominant in the semi-arid Sahelian zone (IFAD, 2006). It is largely subsistence, low-input/low-output, extensive farming based on traditional techniques.[2] Only 15 per cent of Niger's land is arable; millet (*Pennisetum glaucum*), sorghum (*Sorghum bicolor*) and cowpea (*Vigna unguiculata*) are the dominant crops. Average crop yields are low, generally about, 400 kg/ha for millet and 200 kg/ha for sorghum (PPCR, 2010). Small-scale irrigation is of key importance in a number of regions (Tahoua and Maradi). Onions are the most important irrigated crop and large quantities are exported to Nigeria's urban markets. Groundnuts and cotton, which were once important export crops, only contribute marginally to today's economy (FAO, 2005).

The links between poverty, vulnerability, land degradation and low agricultural productivity are strong in rural Niger. According to the Ministère de l'Eau, de l'Environnement et de la Lutte Contre la Désertification (2011), the average yields of major crops have not increased significantly over the last three decades.[3] Since the 1980s, food production per capita has remained stagnant, with strong spatial-temporal variability. As a result, Niger is regularly exposed to food insecurity and must resort to imports and international aid (IFAD, 2007). This situation can be attributed to a number of factors which limit agriculture productivity: climatic constraints (in particular rainfall variability and persistent droughts) and an overall trend towards a drier climate (Figure 7.1), land degradation, the prevalence of sandy soils with limited potential, strong demographic pressure,[4] limited access to inputs and equipment that would favour intensification, as well as serious inadequacies in basic services and infrastructure (IFAD, 2007).

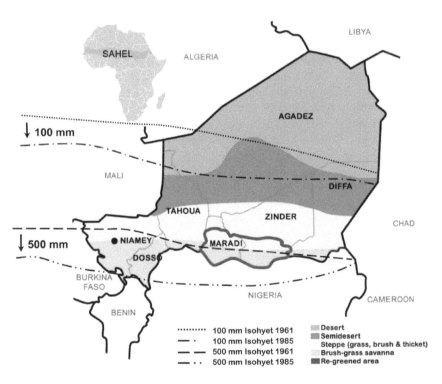

Figure 7.1 Regional boundaries and vegetation zones in Niger. The hatched lines show the southward shift of the 100 mm and 500 mm rainfall isohyets during the drying trend from 1970 to the late 1980s (Koechlin, 1997: 17). The boundaries between vegetation zones are more gradual shifts in species distributions than sharp, distinct borders. The bold line encircling Maradi shows an estimate of the area regreened in Maradi and Zinder regions of south–central Niger (Source: Sendzimir et al., 2011)

Irrigated agriculture, producing mainly onions, but also rice contributes about 14 per cent of the total agricultural production (FAO, 2005). The variety of irrigated crops has increased over time. In 2005, FAO estimated that the total irrigated area was about 73,600 ha (1.6 per cent of the cultivated area). The relatively low rate of annual increase in irrigated area (0.6 per cent over the period 1989–2005) is partly explained by the high costs of irrigation development. It ranges between US$8700 and 14,000 per hectare, depending on the size and specific characteristics of the sites and the systems used for water diversion (FAO, 2005). The total irrigated area may, however, be underestimated as traditional irrigation based on hand-dug wells plays an important role, for instance in parts of the Tahoua region.

In the last three decades, national policies and donor programmes have given greater attention to water harvesting – considered to be a more viable, long-lasting and more appropriate solution to food insecurity. In 2005, the area under water harvesting ('*superficie en collecte des eaux de ruissellement: conservation des eaux et des sols, défense et restauration des sols CES/DRS*') was estimated to be about 300,000 ha and the cost for the development of water harvesting structures ranged between US$30 and 400 per hectare depending on the technology or combination of technologies (FAO, 2005). In 2009, the World Bank published a study on the impact of sustainable land management programmes on poverty in Niger (World Bank, 2009). The study took into account 31 of these and calculated that, from the early 1980s, about 207 billion CFA (approximately US$450 million in 2009) were spent in programmes having soil and water conservation, land rehabilitation and tree planting components (Ministère de l'Eau, de l'Environnement et de la Lutte Contre la Désertification, 2011). Besides this, farmers have invested themselves in simple water harvesting techniques, but also in the development of new agroforestry systems. Since the middle of the 1980s farmers, in the densely populated parts of the Maradi and Zinder regions, have protected and managed the spontaneous regeneration of woody species on their farmland. This has led to the construction of new agroforestry parkland on about five million hectares (Reij et al., 2009) and overall 'regreening' (Figure 7.1). New agroforestry systems have also emerged on land treated with water harvesting techniques. Farmers use manure in the planting pits and the semi-circular bunds. The manure contains seeds from trees and bushes, which benefit from the combination of water and fertility. They protect and manage species, including *Piliostigma reticulatum, Combretum glutinosum, Faidherbia albida, Guiéra senegalensis* which they consider useful as sources of firewood, fodder and/or for improving soil fertility.

The evolution of rural development policies since Niger's independence can help explain the increased attention and investments in water harvesting. In particular, since the major drought of 1968–73, the different governments have emphasized the need for self-sufficiency in food production, which explains the significant investments in water harvesting since the middle of the 1980s, but also Niger's attempts to develop a land tenure policy. Until the early 1980s Niger was relatively rich compared to its neighbours as it benefited from the export of uranium. In 1986, the nuclear disaster at Chernobyl reduced investments in nuclear energy and demand for Niger's uranium dropped dramatically, which led to a steep fall in government income. In 1987 President Kountché died and this marked the beginning of a long period of economic crisis and political instability. It is ironic to note that despite the economic and political crisis, this is the period when all the major investments in water harvesting for land restoration took place. The current President, Issoufou Mahamadou, who was democratically elected in 2011, has made food self-sufficiency a pillar of his development policies as shown by the recently launched programme '*Nigériens Nourrissent Nigériens*' (Nigerians feed Nigerians).

History and importance of water harvesting practices

Water harvesting bunds and trenches[5]

Water harvesting bunds (*diguettes* in French) and trenches (*tranchées* in French) were introduced in the Keita Valley, Tahoua Department, of Niger by the Keita Valley Integrated Development Project implemented by FAO (1984–96). These water harvesting techniques were used as tools in rehabilitating degraded land and were devised for growing crops, fodder and for afforestation.[6] The structures were made either by hand, using food-for-work (FFW), or by a combination of machinery and hand labour (Photos 7.1 and 7.2).

The Keita Valley project used machinery to construct bunds on the plateaux (see video footage of the construction sequence in Critchley and Reij, 1988). These bunds were 50 cm high and 75–100 m long. Each bund had upward sloping wings on both sides with a length of about 15 m. The bunds were stone-pitched, by hand on the top and backslope in order to stabilize them and reduce maintenance requirements. The area impounded by each bund was planted. On slopes of less than 0.5 per cent, the spacing between the bunds was 45 m, thus allowing for an unplanted catchment strip of 30 m between cultivated plots, giving a catchment: cultivated area ratio of 2:1. A small gap of 3–4 m was left between individual bunds in the same rank to allow passage of overflow. Crops were planted in the impounded area while trees were planted along the bunds at a distance of about 5 m apart. The land was ploughed for the first two seasons by tractor to break the crust in order to allow rainfall and runoff to infiltrate. Millet was the most common crop and it was planted in furrows. The preferred tree species for planting along the bunds were *Acacia seyal*, *A. nilotica* and the exotic *Prosopis juliflora*. In 1987, the estimated cost for establishment of the bunds was about US$574 per hectare (Critchley et al., 1992).

Bunds were not only constructed on the fairly level plateaux, but also on gentle slopes between 1 and 3 per cent (*glacis* in French) and at slightly closer spacing. Their purpose was to harvest and spread runoff from a larger, external catchment. The main difference with the other technique was that the whole area was planted, as it received an external supply of runoff. Also in this case, bunds were

Photos 7.1 and 7.2 Bunders and compactors used on constructing bunds in the 1980s
(W. Critchley)

covered all-over with stone for reinforcement. The most common crops grown were sorghum and millet. The land was ploughed by tractor for the initial seasons. Trees such as *Acacia seyal*, and the exotics *Prosopis juliflora* and *Parkinsonia aculeata* were planted in front of the bunds.

The Keita Valley project introduced also hand-dug trenches, though at an early stage they were dug by a tractor-mounted digger. They were promoted as a suitable technique for tree planting on hillsides where there was availability of stones. Each trench was designed to hold about one cubic metre of water (dimensions: 3 m long, 60 cm wide, 60 cm deep) and was served by a catchment area of 12.6 m². About 700 trenches were constructed per hectare (Critchley et al., 1992). To reduce the potential danger of waterlogging, a 20-cm-high step was left in the centre of the pit on which the seedling was planted – usually in the month of August when runoff from the early rains had already been stored. The preferred tree species were *Acacia nilotica, A. seyal, A. radiana* and *Prosopis juliflora*. By 1987, about 150,000 trenches were constructed (equivalent to over 200 ha) and the total cost was about US$1750 per hectare (Critchley et al., 1992).

Between 1984 and 2003, the Keita project rehabilitated about 34,000 ha using a mix of water-harvesting techniques and sand dune fixation (Di Vecchia et al., 2007). The techniques are technically effective in rehabilitating degraded land. However, Critchley et al. (1992) concluded that the techniques used by the Keita project were not only expensive but their adoption was tied to the availability of machinery and project support. Even adequate maintenance of these techniques was likely to be very limited, given the size and cost of the structures.

Demi-lunes (semi-circular bunds)

Demi-lunes as a water harvesting technique had never been used traditionally in Niger or elsewhere in the Sahel as far as was known (Critchley and Siegert, 1991). Its historical development in Niger dates back to the mid-1970s. Semi-circular bunds or *demi-lunes* were first introduced into Ourihamiza at the initiative of the Catholic Mission of Tahoua as a measure to counteract the drought of 1973 (Rochette, 1989). From that year, construction of *demi-lunes* became accepted as a recommendation by the Regional Development Council. In 1985 the total area covered by this system was estimated at about 400 ha, but the number of hectares treated increased over the years as a result of the gradual introduction of the technology into Tahoua region and the rest of the country. In 1988, *demi-lunes* were introduced in the district of Illéla as part of the IFAD-funded Soil and Water Conservation programme Programme Spécial National FIDA NIGER Phase 1 (PSN1). One of the sub-projects was based in Badaguichiri. The implementation of the project was preceded by an awareness campaign. The traditional work parties, termed *gaya* in Hausa, were used to mobilize the local people for the construction of the *demi-lunes* in both collective and individual fields. Training was given to the technical committees responsible for monitoring the work and tools were made available in the targeted villages.

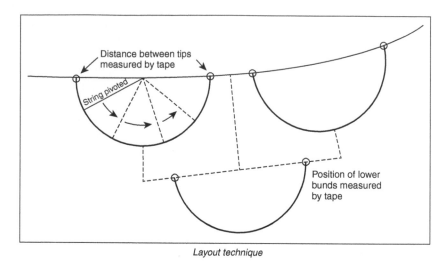

Layout technique

Figure 7.2 Demi-lunes (Source: Critchley and Siegert 1991, p. 65)

The semi-circular bunds in Ourihamiza were documented by Rochette in 1989 and by Critchley et al. in 1992. Both studies provided a detailed description of the application of this water-harvesting technique for crop production (millet and sorghum) instead of the most common use for tree growing and fodder production.[7] The system was described as an in-field, short slope or microcatchment technique, consisting of small semi-circular structures with radii of only 2 m, tips of the bunds on the contour and a staggered layout – thus allowing the collection of runoff from the area between the bunds above (Figure 7.2).[8] The recommended design density was 313 *demi-lunes* per hectare giving a catchment: cropped area (C:CA) ratio of 4:1 (Critchley et al., 1992). The crop was planted only within the *demi-lunes* while the catchment was kept weed free.

In Ourihamiza, *demi-lunes* were constructed by hand labour under a food-for-work arrangement, and the emphasis was on rehabilitation of degraded land in an area where impoverished agropastoralists lived. The average rate of implementation was eight *demi-lunes* per person per day under the cash-for-work (CFW) programme but dropped to four *demi-lunes* per person per day under food-for-work (FFW). In 1987 the cost of constructing *demi-lunes* was estimated to be about US$150 per hectare. This amount included the value of FFW rations, but not the costs for depreciation of tools used (Desbos et al., 1987). Critchley and colleagues (1992), who carried out their study at the end of the 1980s, underlined that even if this technique was relatively simple, there was no spontaneous adoption at that time by the local people. Their hypothesis was that *demi-lunes* were perceived to be too labour-intensive for both construction and maintenance and differed from the usual production practices on sandy soils. They were to be proved partially, at least, incorrect in this hypothesis as will be demonstrated later in this chapter.

The IFAD-funded Soil and Water Conservation programme in the Illéla district (PSN1) adopted the same design as the Ourihamiza project but also tested a design with a lower density of structures per hectare in order to increase the catchment: cultivated area ratio and, in turn, augment the runoff supply to plants in years of low rainfall (Critchley et al., 1992). The semi-circular bunds constructed in Badaguichiri under the IFAD-funded programme were part of the study carried out by Hassane et al., published in 2000 under the title 'Water Harvesting, Land Rehabilitation and Household Food Security in Niger'. At the time of the original study, *demi-lunes* were constructed manually and collectively (Critchley et al., 1992). Labour was provided by local people (mainly Hausa agropastoralists) in the framework of CFW or FFW programmes. The need for collective involvement in the construction phase was justified by the training and extension services provided in land identified by the farmers themselves.

Planting pits (tassa)

In Niger, *tassa* planting pits were a traditional technique sporadically used to rehabilitate degraded plateaux (Hassane et al., 2000). This microcatchment technology was rediscovered in 1989 during a farmer exchange visit to Burkina Faso organized by the aforementioned IFAD programme.

The IFAD programme aimed to increase food security by means of sustainable production achieved through 'small-scale, simple and replicable' conservation practices which could be promoted by local staff and applied by individual farmers. However, it is important to note that in its first year the project had some difficulties in putting its main objective into practice. The national government staff involved had no experience with simple conservation practices and so started off doing what was standard practice for all soil and water conservation/water harvesting (SWC/WH) projects in this region. They hired a grader and a bulldozer for bund construction and for ripping the barren crusted land (Hassane et al., 2000).

After a difficult start-up, the adoption and reproduction of practices by local farmers became a key factor in achieving long-term sustained results. Farmer participation was strongly promoted; local people were involved in the decision-making process from the beginning. Initially the IFAD programme did not focus specifically on *tassa,* but on the construction of contour bunds (2300 ha in four years), and on the development of *demi-lunes* (320 ha). The course changed in the second year of the programme (1989). Exchange visits became a key activity and were much appreciated by farmers. It was during an exchange visit to Yatenga in Burkina Faso that a group of farmers from Illéla district were shown contour stone bunds and improved traditional planting pits (*zaï*) which were the most successful conservations practices in the region. The *zaï* reminded them of the *tassa* technique used in the past in their own region, but long abandoned.

This farmer study visit to Burkina Faso triggered a spectacular process of dissemination of improved planting pits in Niger's Tahoua Department – and from 1989 the project actively stimulated this process (Hassane et al., 2000). However incentives proved a thorny problem:

> The project struggled from the start with what kind of incentives to offer to farmers. All SWC/WH projects in the region offered food-for-work rations to farmers involved in bund construction and other conservation activities. This had been the standard incentive for almost two decades. The IFAD project adopted as its policy that food-for-work would only be provided in years of drought to villages where the harvest had failed. In years of normal rainfall collective SWC/WH efforts could be remunerated with community infrastructure, such as classrooms or village wells. Tools-for-work was also a popular form of support. They were lent to village groups to undertake work collectively.
>
> (Hassane et al., 2000)

The original *tassa* pits were of small size (10 cm diameter and 5 cm depth), the excavated soil was not necessarily put downslope of the pits and no manure/organic matter was added (IFAD, 1992). The pits were made with a hoe to break the surface crust on existing farm fields before the onset of the rains. On the basis of their observations in the Yatenga region of Burkina Faso, some farmers decided to improve their traditional pits by increasing the dimensions and adding manure (IFAD, 1992). Typically, improved *tassa* would have a 20–30 cm diameter, a 10–25 cm depth and would be spaced about one 0.8 m apart in each direction.[9] Farmers would dig approximately 14,000 pits per hectare. The excavated soil would also be placed downslope of the planting pit to allow runoff to enter and hold it better. Organic matter (a mix of manure and grasses) would be added to improve soil fertility.[10] *Tassa* would be dug by hoe in the dry season and at the beginning of the wet season, millet or sorghum would be sown in the pits. Preferably every year, between March and May, sand would be removed from the *tassa*. Every second year the pits would be enriched with manure. The formation of new *tassa* could be completed in one year (WOCAT, 2007).

The IFAD project supported farmers by offering training but also some incentives such as hand tools, community infrastructure and the means to transport stones needed for the stone bunds. The total value of incentives provided by the project for collective soil and water conservation was less than US$0.5 million over a period of seven years. The IFAD project introduced also small experimental plots where not only manure was applied to the pits, but also some inorganic fertilizers. In 1989, three hectares were treated with improved *tassa*: this rose to 78 hectares in 1990. Significantly, and perhaps fortuitously in some respects, 1990 turned out to be a drought year and only farmers who had treated their land with *tassa* or *demi-lunes* were able to harvest. This led to an exponential increase in the adoption of *tassa* (IFAD, 1992). According to Hassane et al. (2000), project statistics indicate that by the end of 1995 about 3800 ha had been treated with *tassa* in the Illéla district alone.[11] But these statistics were only based on what extension agents had been able to measure, and therefore the data may have been conservative. Field observations indicated that *tassa* were increasingly being adopted elsewhere in the Tahoua Department but no figures were available from other districts (Hassane et al., 2000).

Key water harvesting developments: the technologies and the approaches

Water harvesting bunds and trenches

The Keita Valley Integrated Development Project, through which the water harvesting bunds and trenches were first introduced, ran full steam ahead from 1984 to 1996. The project had several follow up phases that extended the total period of intervention up to 2006 (over 20 years in total). By the end of 1999, the project had applied soil and water conservation techniques on about 10,000 ha, and had planted 16 million seedlings under the reforestation programme (IFAD, 2008). In 2003, these figures went up to about 34,500 ha under reclamation and improvement of agricultural and pasture lands, reforestation and dune fixation and 18 million trees planted (Di Vecchia et al., 2007). However, after about 15 years, activities slowed down considerably.

During the Keita Valley project, the construction of bunds and trenches was promoted and supported with a diverse range of incentives such as FFW, provision of inputs (seeds and fertilizers) and equipment, training, and construction of community infrastructure (classrooms, wells, cereal banks, rural roads and community-managed village shops). The implementation was under the supervision of a technical team of agricultural engineers. It required the use of heavy machinery and manual labour provided by the local people. Village committees were established to ensure the proper management of the reclaimed land and the maintenance of the structures. Numerous positive impacts were attributed to these techniques, in particular the decrease of degraded land and the associated increase in agricultural area (about 4800 ha rehabilitated during the first six years of the project), the improvement of crop yields (above the average of 300–350 kg/ha for the area), the increased availability of fodder, the reforestation of degraded land, the reduction of runoff and siltation of the valley bottoms, improved infiltration of rainfall and increased groundwater recharge. In addition to the agronomic and environmental impacts, bunds and trenches had also important socio-economic impacts in terms of improved food security, livelihood diversification (e.g. livestock-related activities) and reduced outmigration especially of young and more vulnerable people.

The project area and the spread of the promoted water-harvesting techniques were revisited by a team of local researchers in 2012; about 25 years after the World Bank's Sub-Saharan Water Harvesting Study (SSWHS). The revisit showed that, as predicted, in Laba and Tamaské (Keita department), bunds and trenches had not been taken up/adopted voluntarily neither by individual farmers nor communities. The basic problem was not their performance but, especially for the bunds, the farmers' inability to construct these without significant external assistance. Over the years, these technologies have been introduced in other regions through state and donor-funded projects in the area of natural resource management. However, in the original project areas, while the majority of the structures have survived a long time, lack of maintenance is slowly but surely taking its toll. The current situation is the outcome of several determinants. First of all, the village committees that were established in 1984 for the maintenance of the structures were operative for approximately three years after the end of the Keita project. Once they ceased to exist, it became clear that these type of structures established with heavy machinery on large areas could not be maintained manually without outside incentives. Secondly, the Keita project overlooked the traditional land tenure system. The project neglected the tenure status of the degraded lands on which it tried to restore productivity. The treated lands were then redistributed to village members without adequately taking into account the rights of the original land owners.[12] Thirdly, the project had a technical approach which disregarded important contextual factors related to people's poverty and vulnerability. Finally, it turned out that the higher the distance of the rehabilitated fields from a village, the poorer the maintenance of the bunds.

Demi-lunes (semi-circular bunds)

Soon after the beginning of the IFAD-funded Soil and Water Conservation programme (PSN1) in 1988, the construction in communal fields was abandoned due to the lack of maintenance. The focus then shifted entirely to land rehabilitation of individual farmers' fields and on degraded land to which individual farmers had clear land rights. The project trained farmers to construct the *demi-lunes* and provided them with tools to do so, but after the project ended farmers inevitably had to rely more on their own means. This approach, in contrast with the very dominant FFW concept promoted by all other SWC/WH projects in the region, allowed the voluntary adoption of the technology even after the programme ended in 1996. A second phase of the programme (PSN2), was supposed to run from 1998 to 2004 with the objective of integrating the lessons learnt from PSN1 in more comprehensive village land management plans. However, due to various procedural and financial constraints, the project had to be suspended. After 1996, project support for the promotion of *demi-lunes* in this region dwindled; most projects focused on investment in water spreading (Federal Ministry for Economic Cooperation and Development-BMZ, undated).[13] Nevertheless, *demi-lunes* continued to be used by farmers, who sometimes expanded the area covered by them.

One of the ways in which low-income farmers managed to rehabilitate degraded land with semi-circular bunds was through the organization of traditional work parties (*gaya*). A farmer can invite members of the same age group and ask them to help him with the construction of *demi-lunes*. In return, he provides food and drinks to the workers, who accepted the invitation and he is expected to respond to similar invitations from other farmers. The introduction of *demi-lunes* seems to have increased interest in the organization of such traditional work parties. Farmers also show a preference towards simpler water harvesting technologies such as planting pits (*tassa*) to restore degraded land to productivity and to improve soil fertility. Other projects (also through NGOs) have been set up in the original sites of the Keita Valley project to consolidate its achievements.

The project area and the spread of *demi-lunes* were revisited in 2012 about 12 years after the study by Hassane et al. The revisit showed that, in Badaguichiri (Illéla district), *demi-lunes* have now become part of farmers' production practices. In several villages, like Farabani, Toumboul, and Intouramé, most of the cultivated land is now treated with small *demi-lunes*. Farmers have reduced their size and their spacing. They have reduced the catchment area, which means that the technique no longer functions as a water-harvesting technique, which collects and concentrates runoff from an uncultivated area, but functions as an *in situ* soil and water conservation technique. This is, however not a completely new phenomenon because such systems were already documented in 1989 by Rochette.

In the absence of a detailed survey at regional level, it is not possible to estimate the scale at which *demi-lunes* have been adopted, but observations indicate that farmers are continuing to maintain, and gradually expand, them.

Planting pits (tassa)

Tassa/zaï planting pits, in combination with stone bunds, are the biggest success story of water harvesting in West Africa over the last 25 years (Critchley, 2009).[14] This system has been extensively studied and written about (Anschütz et al., 1998; Hassane et al., 2000; IFAD, 1992; Liniger et al., 2011; Reij, 1991; WOCAT, 2007).

The main reason for the rapid adoption of improved *tassa* is the fact that the technique is simple and produces results the same year. Farmers can dig *tassa* incrementally depending on how much labour they are able and willing to invest. *Tassa* can be applied by individual farmers who treat just their own fields to which they have land use rights. As a result, very few land tenure problems arise. Disadvantages appear in the case of rehabilitation of communal land. Conflicts arise mainly between farmers and pastoralists when pasture land is being turned into cultivated fields. Farmers appreciate that with this technique, little land is lost to the structures and fields treated with improved *tassa* produce some yields even in years of low rainfall. By concentrating runoff water, they can tide a crop over a drought spell.

Tassa have proven to be technically effective in capturing and holding rainfall and runoff while, simultaneously, improving water infiltration and nutrient availability (WOCAT, 2007). But the question is how farmers perceive this impact on yields. Farmers tend to compare the before situation (barren degraded land with no yield) with the situation after their investment (average yield of 500 kg/ha) and

Table 7.1 Impact of *tassa* and *demi-lunes* on millet yields, on-farm: 1991–1996 (kg/ha)

Rainfall Badaguichiri Illéla	1991 726 mm 581 mm	1992 423 mm 440 mm	1993 369 mm 233 mm	1994 613 mm 581 mm	1995 415 mm 404 mm	1996 439 mm 440 mm	Average 1991–1996
Tassa							
T0	0	125	144	296	50	11	125
T1	520	297	393	969	347	553	513
T2	764	494	659	1486	534	653	765
Demi-lunes							
T0	0	86	77	206	28	164	112
T1	655	293	416	912	424	511	535
T2	1183	538	641	1531	615	632	857
Average farmers' yields Illéla	386	241	270	362	267	282	301

(Source: Hassane et al., 2000) Legend: T0 = without situation (control); T1 = SWC/WH technique + manure; T2 = SWC/WH technique + manure + fertilizer

an ability to harvest some crop in drought years (Table 7.1). This explains why farmers still continue to invest in *tassa*.

This trial illustrated that using water-harvesting techniques in this situation without soil fertility management produced poor crop yields. When manure is added (T1) the average crop yields for *tassa* range from 297 kg/ha to 969 kg/ha with an average of 513 kg/ha over a six-year period. The average yields for *demi-lunes* are broadly similar. When some inorganic fertilizers (T2) are added, this yields another 250–300 kg/ha.

In terms of costs compared to benefits, land users are positive about the establishment as well as the maintenance over the short-term. In the long run the results are even slightly better, as land users consider both to be very positive (WOCAT, 2007).[15]

Tassa planting pits, as well as *demi-lunes*, not only have a positive impact on yields and food security, they also lead to the development of new agroforestry systems. Whereas most plateaux in the Illéla district were barren in the 1980s, basically all fields that were treated with *tassa* and *demi-lunes* now have stands of *Piliostigma reticulatum*, which produces fodder for livestock and is perceived to have a positive impact on soil fertility, or with *Combretum glutinosum*, which produces good quality firewood (Photo 7.3).

According to anecdotal evidence, these water-harvesting technologies of *tassa* and *demi-lunes* have also had a significant impact on the water level in wells. As for the impact on groundwater recharge, no formal studies have been carried out and gains are based on people's perception. According to the villagers of Batodi the depth of the water in their wells was 18 m in 1994, but in 2004 the

Photo 7.3 Land rehabilitated with tassa in 1989 (C. Reij, Sep. 2006)

water level had risen by about 14 m and they had created four dry-season vegetable gardens around wells. In January 2004, the water level in their wells had remained high and the number of vegetable gardens had more than doubled. They were cultivating onion on what used to be barren plateau. When asked to explain the increase in water levels in their wells, they answered that it was 'due to Allah', but they acknowledged that before they had treated their land with water-harvesting techniques, almost all rainfall ran off their fields, whereas now almost all rainfall infiltrates.

The rapid and tangible gains produced by *tassa* have contributed to the emergence of flourishing labour and land markets. Farmers rely on hired labour or on *gaya* for the construction of *tassa*. This new source of cash income means that resource-poor farmers don't have to sell their livestock in case of a poor harvest and can avoid seasonal or longer-term migration (Hassane et al., 2000). A small land market already existed in the 1960s and 1970s, but the large scale adoption of *tassa* boosted this market. Since the early 1990s, in many parts of the Tahoua Department, farmers and traders have started buying and selling the *fako* (degraded land) to rehabilitate and make it productive. In 1998, 40 per cent of a sample of 79 farm household heads responded that they had plots of degraded land and the costs of such land had doubled or even quadrupled over a period of ten years (Hassane et al., 2000). No information is available to assess the impact of this emerging land market on the distribution of land ownership. Hassane et al. (2000) put forward the hypothesis that a small rural elite (relatively rich farmers, traders, religious leaders) was buying land, whereas the relatively poor farmers sold land more quickly in case of failed harvests or other calamities.

Conclusion

In a context like Niger where population pressure on available land resources is very high, simple water-harvesting techniques have allowed a considerable expansion of the resource base through the rehabilitation of degraded land. After about 23 years from their introduction in the Illéla district, *demi-lunes* and *tassa* planting pits continue to be used by farmers to restore degraded land to productivity. This is despite the fact that the related project activities stopped 16 years ago. Because it is carried out on a voluntary basis, without outside support, it is a vindication of the water harvesting approaches that introduced these technologies. The area has witnessed a progressive integration of these water-harvesting technologies into the agricultural system and consequently seen a remarkable environmental transformation. Each farmer has individually treated small plots of land, but taken together thousands of hectares of degraded land have been restored to productivity not only in the Illéla district, but also in other parts of Tahoua region and even in other parts of Niger.

The multiple impacts of *tassa* and *demi-lunes* are important to farmers. They have a positive impact on crop yields, but also promote the emergence of new agroforestry parklands, which are beneficial to soil fertility and increase the availability of firewood and fodder. Their impact on local groundwater levels is also generally positive but evidence is anecdotal and non-systematic. Such impacts have been found in some villages, but not in others. There may very well be multiple beneficial impacts, which if expressed in economic terms, could explain how rational it is for farmers to invest in simple water-harvesting techniques such as *tassa* and *demi-lunes*. However what is clear is the paucity of data to match these observations, however obvious they may seem. This confirms a common concern regarding water harvesting in Sub-Saharan Africa: the need to engage in regular monitoring, combined with specific scientific studies, to verify basic parameters including extent of practices, performance of crops, environmental effects and impacts on livelihoods.

As for the bunds and trenches in the Keita department, voluntary adoption was never likely to occur due to the substantial external assistance needed to establish them. This was predicted under the Sub-Saharan Water Harvesting Study (Critchley et al., 1992). Therefore, despite the relative good *technical* performance of water harvesting bunds and trenches the very high costs – and partial dependence on machinery – rule them out of anything but a purely outsider-driven aid programme. There is a social problem also: farmers use existing bunds only if located in the proximity of the village and in lands which are not under dispute. The Keita Valley Integrated Development Project, through its technologies, may have contributed to recovering and enhancing the communal natural resources in the area but failed in establishing a formula for long-term management of these resources. The bottom line, as stated by Di Vecchia and colleagues (2007) is that land reclamation and conservation practices for agricultural activities are very important, but not sufficient in themselves. The Keita experience has proved that the recovery period takes twice as long as the degradation process – and that prevention is far less costly than reclamation (Di Vecchia et al., 2007).

Notes

1 This section has been largely compiled from three sources, principally FAO (2005), but also IFAD (2007) and Ministere de l'Eau, de l'environnement et de la Lutte Contre la Desertification (2011).
2 The average farm size under rainfed farming systems is 5 ha, while under irrigation ranges between 0.25 to 0.5 ha per family (Secretariat Permanent de la SRP, 2003).
3 For 14 years during the period of 1983 to 2000, there was a deficit in cereal production. The exceptions were: 1988, 1998 and 1999 (FAO, 2005).
4 The frequency of farmer–herder conflicts is rising as steadily declining rainfall pushes the herders increasingly southwards and as rising population pressures push crop farmers into grazing areas (IFAD, 2006).
5 This section has been largely compiled from Critchley et al. (1992).
6 In the Keita Valley Integrated Development Project the term water harvesting (or its translation) was not specifically used.
7 Quote from Critchley and Siegert (1991): 'Semi-circular bunds, of varying dimensions, are used mainly for rangeland rehabilitation or fodder production. This technique is also useful for growing trees and shrubs and, in some cases, has been used for growing crops.' (p. 59); Quote from Critchley et al. (1992): 'The use of demi-lunes for crop production is unusual, and in this respect the Ourihamiza project is rather unique in Sub-Saharan Africa.' (p. 87)
8 Semi-circular bunds can be designed to a variety of dimensions. They can range from small structures closely spaced (suitable for the relatively 'wetter' semi-arid areas with low slopes and even terrain) to larger and wider spaced bunds (suitable for drier areas and less even terrain) (Critchley and Siegert, 1991).
9 *Tassa* planting pits are usually 15–20 cm deep, and the diameter is 30 cm. The usual spacing is about 90 cm. *Tassa* pits are suited to low slopes (below 2 per cent), high runoff, and hand labour. Construction and maintenance of *tassa* pits is relatively easy but labour-intensive.
10 Manure (if available) would be added, either during October/November or in March/May, with ideally about 250 gram per pit (2.5t/ha). According to Hassane et al. (2000) the quantity of manure per hectare used in the IFAD programme was between 5–6 tons every second year (500 grams per pit). Manure attracts termites that digest it and make nutrients better available to plants. In turn, the termites' channels increase the infiltration of water into the soil.
11 In 1993 the project carried out a survey in 27 villages to get a better idea on the level of adoption. The survey indicated that 46 per cent of households interviewed had applied the improved *tassa* technique (3558 families were interviewed, 1666 families applied *tassa*). A smaller survey was carried out in 1998 in 17 villages. Of the 88 farmers interviewed 84 had invested in *tassa*, 49 farmers combined the application of *tassa* with stone bunds (Hassane et al., 2000).
12 Quote from Rochette, R. (1989): 'The farmers in the village of Laba do not want the project to treat their cultivated land on the plateau because they fear they will lose their rights on it. It is unfortunate that cultivated land is under threat of degradation when adjoin land is rehabilitated at high price' (p. 315).

13 In Niger the use of water-spreading weirs began in the Tahoua region in 1997. Thanks to the weirs, the fertile but heavily damaged valleys were also rehabilitated in addition to the plateaus and slopes, thus stabilizing the drainage basins in their entirety (Federal Ministry for Economic Cooperation and Development-BMZ, undated).

14 *Tassa* planting pits and stone bunds are often combined by the farmers because stone bunds protect the pits against damage by external runoff from an outside catchment. Their combination allows a versatile crop production system in a wide variety of situations in dry areas.

15 Total establishment costs for *tassa* per hectare is approximately US$160. Establishment costs for the labour work to dig *tassa* for one hectare (100 person days) are US$150. The costs for tools to dig the *tassa* is about US$5, and also US$5 for 2.5 tons of manure (WOCAT, 2007). Total maintenance costs for tassa per hectare per year are on average US$ 33.50.

References

Critchley, W. (2009) 'Soil and Water Management Techniques in Rainfed Agriculture; State of the Art and Prospects for the Future', Background note prepared for the World Bank, Washington D.C.

Critchley, W. and Siegert, K. (1991) 'Water Harvesting – A manual for the design and Construction of Water Harvesting Schemes for Plant Production', FAO.

Critchley, W., Reij, C. and Seznec, A., (1992) 'Water Harvesting for Plant Production. Volume II: Case Studies and Conclusions for Sub-Saharan Africa', World Bank Technical Paper Number 157, Africa Technical Department, World Bank.

Desbos, P., Mounkaila, A., Akotey, A., Djibo, H., Deriaz, D., Deriaz Uwantege, E. and Monimart, M. (1987) Expérience no. 2, Ourihamiza/Tahoua – Niger: Demi-lunes, barrages seuils, agroforesterie.

Di Vecchia, A., Pini, G., Sorani, F. and Tarchiani, V. (2007) 'Keita, Niger – The impact on environment and livelihood of 20 years fight against desertification', Working paper no. 26.

FAO (2005) *Irrigation in Africa in figures*, AQUASTAT Survey – 2005, FAO Water Report 29.

Federal Ministry for Economic Cooperation and Development-BMZ (undated) 'Water-spreading weirs for the development of degraded dry river valleys', GIZ.

Hassane, A., Martin, P. and Reij, C. (2000) *Water Harvesting, Land Rehabilitation and Household Food Security in Niger*, IFAD, Rome.

IFAD (1992) *Soil and Water Conservation in Sub-Saharan Africa – Towards sustainable production by the rural poor*, IFAD, Rome.

IFAD (2006) *Republic of Niger – Country Strategic Opportunities Paper*, IFAD, Rome.

IFAD (2007) *République du Niger – Programme Spécial National Phase II (PSN-II) – Evaluation Rapport No. 1920-NE*, IFAD, Rome.

IFAD (2008) *Water and the Rural Poor Interventions for improving livelihoods in Sub-Saharan Africa*, IFAD, Rome.

IFAD (2009) *Agricultural and Rural Rehabilitation and Development Initiative – GEF*, IFAD, Rome.

Liniger, H., Mekdashi Studer, R., Hauert, C., and Gurtner, M. (2011) *Sustainable Land Management in Practice – Guidelines and Best Practices for Sub-Saharan Africa*. TerrAfrica, World Overview of Conservation Approaches and Technologies (WOCAT) and Food and Agriculture Organization of the United Nations (FAO). Technical editor: Critchley, W.

Ministere de l'Eau, de l'environnement et de la Lutte Contre la Desertification (2011) 'Cadre Strategique d'Investissement du Niger en matiere de gestion durable de terres (CSIN-GDT)'.

PPCR (Pilot Programme for Climate Resilience) Sub-Committee (2010) 'Strategic Program for Climate Resilience: Niger', PPCR/SC.7/6.

Reij, C., Tappan, G. and Smale, M. (2009) 'Agroenvironmental transformation in the Sahel: Another kind of "Green Revolution"', IFPRI – Discussion Paper 914.

Rochette, R. (1989) *Le Sahel en Lutte Contre la Désertification. Comité Inter Etat DE Lutte Contre la Sécheresse au Sahel*, CILSS/GTZ.

Secretariat Permanent de la SRP (2003) *Strategie de Developpement Rural – Le secteur rural, principal moteur de la croissance éconimique*, SRP.

Sendzimir, J., Reij, C. and Magnuszewski, P. (2011) 'Rebuilding resilience in the Sahel: regreening in the Maradi and Zinder regions of Niger', *Ecology and Society*, vol. 16, no. 3, p. 1.

World Bank (2009) 'Impacts des Programmes de Gestion Durable des Terres sur la Pauvreté au Niger', Rapport no. 48230-NE.

WOCAT (2007) *Where the Land is Greener – Case studies and analysis of soil and water conservation initiatives worldwide*, Liniger H. and Critchley W. (eds), WOCAT, Bern.

Chapter 8

Tanzania
Bright spots and barriers to adoption

Henry Mahoo, Frederick Kahimba, Khamaldin Mutabazi, Siza Tumbo, Filbert Rwehumbiza, Paul Reuben, Boniface Mbilinyi and John Gowing

Introduction

Farmers who live in tropical semi-arid areas have to cope with frequent negative impacts on livelihoods (e.g. food insecurity and economic losses) as a result of drought, soil erosion or flooding. Over the past 100 years, floods have caused 38 per cent of all declared disasters in Tanzania, while droughts caused 33 per cent. Often the floods and droughts occurred in the same semi-arid area and in the same season. The problem is that up to 70 per cent of the rainfall can be 'lost' as runoff, while only a small fraction remains in the soil long enough to be useful. The farmers' lament is '*mvua moja tungevuna*', ('one more rain and we would have harvested') as they commonly believe that one more rainstorm would have delivered a good crop. A clear win–win solution is to convert the damaging runoff into useful soil-moisture storage needed for crop and pasture growth, but until recently policy makers had not recognized this possibility. Policies have been dominated by two contradicting perceptions. First, that the only solutions to livelihood problems in the drought-prone semi-arid areas were drought-resistant crops or irrigation. Second, that the solution to flooding and soil erosion was disposal of 'hazardous' runoff away from crop and range lands. This led to soil and water conservation programmes that focused on water disposal in areas where agriculture and livelihoods are affected more by shortage of water than anything else. A sustained research and communication effort during the 1990s brought about a remarkable change in perception and policy towards rainfall runoff (Court and Young, 2003). This chapter revisits the research site and reflects on that.

Tanzania: background[1]

The climate of Tanzania varies from tropical along the coast to temperate in the highlands. There are two types of seasonal rainfall distribution:

- The unimodal type, where rainfall is usually from October/November to April, found in the central, southern and south-western highlands.
- The bimodal type, comprising two seasons: the short rains (*Vuli*) fall from October to December, while the long rains (*Masika*) fall from March to June.

This type occurs in the coastal belt, the north-eastern highlands and the Lake Victoria basin.

Annual rainfall varies from 500 mm to 1000 mm over most of the country. Rainfed cropping seasons are broadly:

- short rains (*Vuli*) season from September/October to January/February;
- long rains (*Masika*) season from February/March to June/July; and
- a combination of the two (*Musumi*) from November to June (unimodal rainfall).

The total population was 43 million in 2010 (updated from 2002 census), of which 74 per cent is rural. The vast majority of the population lives inland, far away from the coastline. Poverty is concentrated in the rural areas; however, urban poverty has accompanied rapid urbanization. The national poverty rate is about 36 per cent. The agricultural sector continues to have the highest impact on the levels of overall economic growth. Agriculture provides work for 14.7 million people, or 79 per cent of the total economically active population, and 54 per cent of agricultural workers are female. Small-scale subsistence farmers comprise more than 90 per cent of the farming population. The main food crops grown are maize, rice sorghum, millet, wheat, sweet potato, cassava, pulses and bananas. Maize is the dominant crop with a planted area of over 1.5 million hectares (ha), followed by rice with more than 0.5 million ha. In recent years, the country has not been self-sufficient in cereals, but is self-sufficient in non-cereal food crops at the national level. There is a clear difference in the supply capabilities of staple-food crops among the regions.

Irrigation in the form of traditional irrigation schemes goes back hundreds of years. The potential for expansion of irrigated agriculture is enormous and policy from colonial times until recently has focused on:

- the construction of new irrigation estates for parastatal organizations;
- the construction of new modern style schemes to be run by smallholders; and
- the rehabilitation or upgrading of traditional irrigation schemes.

In general, the development of irrigated agriculture has been slow such that the total area equipped for irrigation is less than 10 per cent of potential. The following types of irrigation schemes are distinguished within the national irrigation policy:

- Modern irrigation schemes – these are formally planned and designed schemes with full irrigation facilities and usually a strong element of management by the government or other external agencies. Such schemes have been developed in the regions of Kilimanjaro, Morogoro and Mbeya. All parastatal managed irrigation schemes also fall under this category.

- Traditional irrigation schemes – these have been initiated and operated by the farmers themselves, with no intervention from external agencies. They include schemes based on traditional furrow irrigation for the production of fruit and vegetables in the highlands and simple water diversion schemes on the lowlands for rice.
- Improved traditional irrigation schemes – these are traditional irrigation schemes on which, at some stage, there was intervention by an external agency, such as the construction of new diversion structures as in various schemes in the Usangu lowlands.

The main irrigated crops are rice and maize, accounting for about 50 per cent and 30 per cent of the irrigated areas. Other irrigated crops include beans, vegetables (including onion, tomato and leaf vegetables), bananas and cotton. Private irrigation schemes produce cash crops such as tea, coffee, cashew and sugar cane. In recent years water harvesting schemes have also begun to receive some attention.

History and development of water harvesting practices in Tanzania

Water harvesting technologies and practices

Any attempt to promote sustainable intensification of agriculture in semi-arid areas of Tanzania must tackle the problem of unreliable and highly variable rainfall. The dominant perceptions outlined above can be seen as the explanation for official neglect, until quite recently, of water harvesting as a viable option. Nevertheless, water harvesting techniques are known to have been in use in Tanzania for generations (Gowing et al., 1999); some are traditional indigenous practices while others involve exogenous technologies. Every water harvesting system involves generating runoff from a catchment area (C) and delivering it to a cropped area (CA). A simple classification[2] is based on size ratio (C:CA) and transfer distance between C and CA.

(i) Low ratio(C:CA < 2) and short transfer distance (< 5 m)

Planting pits have been documented as an indigenous practice in various countries, notably Burkina Faso and Mali, where they were reported in the SSWHS (Critchley et al., 1992). In Tanzania they exist also as a traditional practice in the Matengo Highlands of Mbinga district where they are known as 'ngoro' (Temu and Bisanda, 1996). Pits are typically 2 m wide and 30 cm deep. This water harvesting system was documented during the colonial era (Stenhouse, 1944) and was revisited by Willcocks and Gichuki (1996) and by Kato (2001).

(ii) Intermediate ratio (C:CA < 5); transfer distance 5–50 m

A number of water harvesting systems can be identified which collect overland or rill flow and deliver it to an adjacent crop area. The short transfer distance ensures that the system offers high runoff efficiency and that it is situated within the land holding of an individual farmer.

(a) One variant is strip catchment tillage, which involves alternating strips of crops with strips of grass or cover crops. This technique was introduced during the colonial era primarily as an erosion control measure. Its imposition was seen as coercive and systems were abandoned at independence (Kauzeni et al., 1987).
(b) A second variant also involves creation of cross-slope barriers (stone lines or earth bunds) to intercept runoff. There are no reported examples of stone lines in Tanzania, but contour bunds are common. This technique was also introduced during the colonial era primarily as an erosion control measure. Trash lines are also reported as a traditional practice in Tanzania (Thornton, 1980).

(iii) High ratio (C:CA > 5); transfer distance > 50 m

Water harvesting systems in this category may exploit hillslope runoff processes or divert flood flows from ephemeral streams and gullies. The catchment generally lies outside the land holding of the farmer and this separation may mean that runoff is harvested at times when there is no direct rainfall in the cropped area. These systems often involve collective effort amongst a group of farmers for construction and maintenance.

(a) One variant is found extensively in Tanzania and is used primarily for production of rainfed lowland rice in bunded basins. It is believed to have originated in Sukumaland (Lake Victoria basin) where it is known as the '*majaruba*' system (Photo 8.2). It is arguably not a 'traditional' practice, since it was introduced by Asian migrant workers during the colonial era (Shaka et al., 1996; Meertens et al., 1999). In Tanzania 74 per cent of the total rice area is rainfed lowland and the *majaruba* system dominates (Kanyeka et al., 1994). Its rapid adoption and spread without external intervention is quite remarkable and can be seen as indicative of the potential of water harvesting practices. Runoff is collected from stony outcrops and grazing lands in upslope areas with cattle tracks often used as conduits.
(b) A second variant is floodwater diversion in which flows from ephemeral streams and gullies is diverted by an intake structure into a command area which may be shared by a group of farmers (Photo 8.1). This requires social organization to construct and maintain the diversion structure and to control the distribution of water to cropped fields. This water harvesting system is difficult to distinguish from conventional irrigation and it has therefore been the focus of efforts to improve traditional systems such as in the Participatory Irrigation Development Programme (PIDP, see below). The existing cropping area in Tanzania that uses floodwater diversion systems is 8000 ha while the estimated potential has been estimated at over 150,000 ha (IFAD, 2007).

(c) A third variant involves diversion with storage in ponds known locally as *ndiva* (Photo 8.3) (Mbilinyi et al., 2005).

Evolution of policy

The agricultural policies of 1983 and 1997 envisioned agriculture as the engine of economic growth and food security. Both policies identify low agricultural production and productivity to be the major problems associated with agriculture. Policy during the two decades after independence was dominated by irrigation as central to agricultural development under the political ideology of socialism and self-reliance. During this policy window the government supported regional administrations to develop irrigation in various programmes and projects. In 1994, a National Irrigation Development Plan (NIDP) was prepared with the goal of advancing conventional irrigation. Water harvesting did not feature in the irrigation development plans; however, some NGOs and donor projects supported water harvesting particularly in areas where conventional irrigation was not a possibility. Such initiatives include the Participatory Irrigation Development Programme (PIDP) implemented from 1997 to 2005. PIDP worked mainly in the semi-arid areas of central Tanzania to promote floodwater diversion water harvesting. Under PIDP paddy production under water harvesting was reported to have increased from 1–2 t/ha to 3–4 t/ha (PIDP, 2007).

In 2002 the NIDP was reviewed and water harvesting featured explicitly in the new National Irrigation Master Plan (NIMP) as a component of the overall irrigation strategy. NIMP aimed to develop water harvesting-based irrigation on 57,000 ha by 2012 and 68,000 ha by 2017 (URT, 2002; Chiza, 2005). The 2010 Irrigation Policy aimed to supply reliable water to water harvesting-based irrigation schemes through improvement of traditional water-harvesting infrastructure and software; development of small, medium and strategic large-scale water storage structures and/or inter-basin transfers of water for irrigation purposes in an economically efficient, socially acceptable and environmentally responsive manner; and promote beneficiary awareness for the contribution to the improvement of their water-harvesting irrigation scheme infrastructure. In addition, the policy also supported the rehabilitation, remodelling and upgrading of improved water harvesting irrigation schemes on the basis of cost sharing and cost recovery; provided technical facilitation for farmers to strengthen irrigators' organizations for improved management of their irrigation schemes; and supported the implementation of environmental management plans.

Prior to this change, policies were dominated by two contradictory perceptions. First, that the only solutions to livelihood problems in the drought-prone semi-arid areas were irrigation or drought-resistant crops; secondly, that the solution to flooding and soil erosion was disposal of 'hazardous' runoff away from crop and range lands. This led to soil and water conservation programmes that focused on water disposal in areas where agriculture and livelihoods are affected more by shortage of water than anything else. A sustained research and communication effort during the 1990s brought about this remarkable change in perception and policy towards rainfall runoff and water harvesting (Court and Young, 2003).

Water harvesting technology development and promotion

The Western Pare Lowlands (WPLL) in the north-east of Tanzania are classified as having low potential for agriculture. The farmers here, many of whom have migrated to the area from the high-potential uplands in the Pare Mountains, have a strong preference for maize and have resisted attempts to introduce sorghum as a drought-tolerant alternative. Participatory evaluation established that the farmers were aware of the risks, but preferred to adapt their maize cropping systems to alleviate the current production constraints. This area therefore became the focus for a concerted experimentation, demonstration and modelling effort (Photo 8.4) aimed at promoting adoption of water harvesting practices (Gowing et al., 2003; Hatibu et al., 2003). Work began in 1993 with the establishment of trials on an experimental site and later on farmers' fields and continued until 1999. This demonstrated the potential benefits of water harvesting and brought about a marked shift in perceptions. Whereas previously runoff was seen as a hazard, it is now recognized as a valuable resource. In-field water harvesting systems (category (ii) above) were shown to have some benefits, but farmers were more enthusiastic about adopting external catchment water harvesting systems (category (iii) above).

The first floodwater diversion system was introduced at Kifaru village in 1997. The purpose was to demonstrate techniques for harvesting water from gullies and water conservation measures for crop production (Bakari et al., 1998). This allowed farmers to evaluate the performance of floodwater harvesting systems. The second structure was built in Hedaru in 1999 (Lazaro et al., 1999). Maize yields increased to 5.4 t/ha compared to 1.7 t/ha without water harvesting intervention. Further diversion structures were introduced at farmers' fields in Bangalala and Mwembe villages in 2004. The structures diverted water to farms with *fanya juu* terraces (Makurira et al., 2009). These fields were mainly planted with maize. Average maize yields for four seasons from 2006 to 2007 in the two sites increased from 0.6 and 0.9 t/ha to 1.8 and 2.3 t/ha, respectively.

Success stories in early intervention villages justified scaling out of interventions and action research in the Makanya catchment. The Smallholder System Innovations (SSI) programme constructed diversion systems for two farmers in Bangalala village. These were pilot study plots that integrated floodwater diversion with ripping and *fanya juu* terraces to promote more water and retention within the root zone. A participatory action research approach was adopted where the relevant farm owners were actively involved in setting up the research, monitoring and interpretation of results. The plots served as farmer field schools where more farmers learnt about the technologies. During this time (2000–2003) the focus of efforts shifted to dissemination and uptake promotion (Lutkamu et al., 2005).

Photo 8.1 Floodwater diversion in Same District, WPLL. Source: Soil Water Management Research Group, Sokoine University of Agriculture

Photo 8.2 Majaruba system involving hillslope runoff diversion. Source: Soil Water Management Research Group, Sokoine University of Agriculture

Photo 8.3 Micro-dam (*ndiva*). Source: Soil Water Management Research Group, Sokoine University of Agriculture

Photo 8.4 Water harvesting experiment/demonstration site WPLL. Source: John Gowing

It is not easy to estimate the uptake of a range of water harvesting techniques over time in Tanzania given the lack of records. An attempt made by SWMRG (2001) documents the uptake of water harvesting techniques in two locations, namely Mwanga and Same Districts, Kilimanjaro Region (Table 8.1) and Maswa District, Shinyanga Region (Table 8.2).

In Same District, ephemeral stream/gully flow diversions are the only techniques mentioned for 1940. In Mwanga, ephemeral stream/gully flow diversion and deep tillage are the oldest techniques mentioned to have started in 1950s. For Maswa District, there are more techniques identified that started in 1940s (Table 8.2). In all the districts there is an indication that there has been an increase in uptake of all the water harvesting techniques over time with most adoption said to have occurred in the 1990s and the 2000s.

Table 8.1 Households practicing various water harvesting techniques in Same and Mwanga Districts, Kilimanjaro Region (n = 659)

Year	Number of respondents practicing particular water harvesting technique				
	Ephemeral stream/gully flow diversion	*Diversion from rangeland*	*Diversion from roads/stock routes/footpaths*	*Storage ponds (ndiva)*	*Excavated bunded basins/borders*
1940–1960	13	2	0	8	0
1970–1980	51	5	2	38	0
1990–2000	276	30	5	96	44

Source: SWMRG (2001)

For the *ndiva* water harvesting systems major investments were made from the late 1990s to 2000s. NGOs contributed construction materials that were not locally available and communities contributed local materials and labour. These systems are widely adopted in the Makanya catchment (Figure 8.1). The adoption of *ndivas* is higher in the midslopes (25 per cent), followed closely by upland (22 per cent) and far behind in the lowlands at 12 per cent. *Ndivas* allow midslope farmers to access water during the night when upland farmers are asleep and therefore overnight water can be accumulated and stored into their *ndivas* for use during the day.

Water harvesting development in Western Pare Lowlands since 2000 baseline

Success stories

After the improvement of *ndivas* and floodwater diversion infrastructure, some storylines of success were documented during the revisit in 2011. In Hedaru

Table 8.2 Households practicing various water harvesting techniques in Maswa District (n = 701)

Year	Number of respondents practicing particular water harvesting technique				
	Ephemeral stream/gully flow diversion	*Diversion from rangeland*	*Diversion from roads/stock routes/footpaths*	*Storage ponds (ndiva)*	*Excavated bunded basins/borders*
1940–1960	0	11	28	14	4
1970–1980	0	50	163	83	12
1990–2000	3	198	503	208	29

Source: SWMRG (2001)

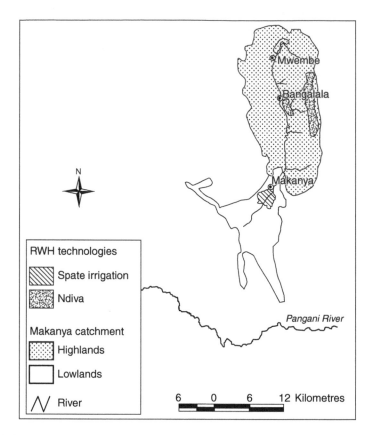

Figure 8.1 Location of water harvesting systems in Makanya catchment

village in 1994 the number of farmers diverting floodwater increased from 200 to about 500 households. A recent survey (2011) found that about 600 ha was under water harvesting irrigation compared to about 200 ha during the 1990s. In Makanya, the area under water harvesting irrigation expands and shrinks seasonally within the potential area of over 700 ha depending on the amount of runoff. Many farmers joined the water harvesting scheme by opening new farms further downstream of the main canals. In Kifaru village the cropland area under water harvesting irrigation expanded from 280 ha in 1998 to 500 ha in 2011. The number of farmers diverting floodwater water into their fields had increased almost four times. During the survey, a farmer who was introduced to water harvesting irrigation in 1998 reported to be still harvesting up to 5 t/ha of maize compared to nothing without the water harvesting technique particularly in bad years. The water harvesting system constructed at Hedaru in 1999 with assistance from Sokoine University of Agriculture (SUA) had partially silted up but was still operational. In Hedaru village, water harvesting caused conversion of grazing land into cropland. At the moment about 600 ha is under this system and managed by smallholder farmers.

The promotion of *ndivas* in the uplands between 1989 and 2000 by NGOs in Mwembe and Bangalala villages was popular (Tumbo et al., 2010). The projects assisted farmers to build and rehabilitate the existing *ndivas*. Due to this assistance and long-term existence of the project (1989 to 2000) more farmers adopted the technology. Between 2002 and 2003, there was an introduction of water user associations governing irrigation water from *ndivas*. This advanced the efficiency of water management and hence increased production of crops (Enfors and Gordon, 2008). The survey carried out in 2011 at Manoo, Ndimka, Kavengele and Makanya areas to know the status of the *ndivas*, identified that the systems are in operation. The Manoo and Ndimka *ndivas* have been improved with NGO support. The Kavengele *ndiva* is unimproved and has not been well maintained allowing high rate of water losses such as seepage and leakage losses.

Reasons for success

One of the reasons that has contributed to acceptability and success of water harvesting systems is crop failures in areas that receive erratic and low seasonal rains (amounting to less than 300 mm per season). Crop production without water harvesting often results in crop failure. With runoff diversion systems, farmers are able to capture water from rains that fall on mountain slopes as much 20 km away from their farmlands.

In Hedaru village several other reasons have contributed to success and sustainability of water harvesting practices. The village receives low annual rains and often experiences flash floods lasting for very short time periods. The floodwater diversion system helps to control and capture the flash floods from mountain areas and divert the water into the fields while minimizing damage to settlements and other infrastructure. Due to low rains in drier years, farmers who grow crops without water harvesting normally experience total crop failure as a result of low moisture in the fields. The areas that have been using the microdams normally receive bimodal rainfall that is low and unreliable, ranging between 500 and 1100 mm per annum, with the highlands receiving more rainfall (Tumbo et al., 2010). This amount of rainfall is not enough to allow rainfed agriculture and for livestock consumption thus necessitating the use of water harvesting systems as adaptive measures. Technical and financial support from external financial institutions (NGOs) for construction and rehabilitation of micro-dam infrastructure are also contributing factors to success.

In the first place, the difference that floodwater diversion and *ndiva* would make in upgrading the productivity of dryland agriculture was the major driver of adoption. Another factor that seems to have helped adoption is the participatory extension and planning interventions implemented by the researchers and other change agents in collaboration with local NGOs.

Problems

Infield management is crucial to facilitate spreading of diverted water and hence the adoption of this water harvesting technology. Absence of infield water spreading techniques made some farmers shy away from floodwater diversion for fear of causing erosion. Commonly used makeshift head works often fail resulting in the whole gully changing its course and thus damaging crop land through erosion. Diversion of runoff water has thus been regarded by some farmers as a potential danger that should be avoided. Others went to the extent of constructing cut-off drains to ensure runoff water did not get into their fields even when rainfall is known to be inadequate. Adoption of *in situ* field water harvesting and spreading technologies including ploughing/tillage and use of small bunded plots or ridges has resulted in increased adoption of spate irrigation. Frequent destruction of the makeshift head works after each flood and therefore the need for reconstruction is one of the challenges under spate irrigation. Changes in the river bed and flash floods carry sediments which are deposited on the fields, raising the elevation of the irrigated land every year and making it increasingly difficult to divert the water into the plots (Komakech et al., 2010). From field observations, siltation is particularly a serious problem in the relatively flat lowlands.

A number of bottlenecks limit the efficiency of the *ndiva* system. Makurira et al. (2007) identified for example that the command area served by any one *ndiva* is often too big compared to the available water. New members are added to the *ndiva* group without addressing the aspect of adequacy of water. Conveyance losses are high due to long distances from the storage structure to the scattered individual farmers' fields (Makurira et al., 2007). All *ndivas* that were visited recently by the authors are still operating, though some were operating below their capacities. It has been reported that during *vuli* fewer households receive water allocations, thus leading to crop failure (Mutiro et al., 2006). This is due to shortage in rainfall. *Ndiva* users fail to cope with the drought because the system distributing runoff water does not have the capacity to buffer meteorological droughts (Enfors and Gordon, 2008). Some of the micro-dams have exceeded a threshold number of users and command area in relation to their capacity (Makurira et al., 2009). Systems that are located further downstream such as the Manoo and Ndimka *ndivas* have more problems of water shortages necessitating negotiations between upstream and downstream water user groups.

Studies by Sokoine University of Agriculture (SUA) on the extent of use of water harvesting technologies in the Lake Victoria and Kilimanjaro regions (SWMRG, 2001, 2005a, 2005b) indicate that the potential has not yet been fully utilized in Tanzania. The critical limiting factors to adoption in the two locations included among others: poor technical knowledge of water harvesting technology, shortage of capital, inappropriate equipment and machinery for constructing water harvesting structures, conflict between crop producers and livestock keepers, weak coordination among users, and siltation of storage structures. In both regions, poor knowledge hampered progress of water harvesting technologies for crop production. The problems are manifested in terms of inappropriate designs and capacity of the structures, poor construction and sometimes lack of flood control embankments. The conflicts between crop producers and livestock keepers reflect a national dimension. The major strategy advocated is village land use planning that has not always been effective.

Investing in improvement and adoption of water harvesting technologies

Experience from previous interventions on water harvesting technologies indicates that improved water harvesting technologies have great potential for increasing crop yields, thereby improving the livelihood of farming communities. In order for the improved technologies to be sustainable there is a need for ensuring participatory approaches in the design and implementation of the technologies. Farmers should contribute a certain percentage of the costs in order to create a sense of ownership. In addition, service charges should be introduced to users to assist in infrastructure rehabilitation; and user groups need to be formed with committees that report to the village government. Table 8.3 presents possible improvements on floodwater diversion and micro-dam technologies within the Makanya catchment.

Interviews with key informants in the study area indicated that supporting knowledge transfer and promotion of adoption of appropriate water harvesting technologies and practices require among other things, implementation of farmer field schools (FFS), use of demonstration plots and farmer field days, farmer education on water conservation measures, farmers exchange visits, and frequent training on proper operation and maintenance of the technologies.

Success with adoption and upscaling of the water harvesting technologies requires the local government authority and village governments to intervene by enforcing laws that restrict environmental degradation thereby ensuring sustainability of the improved water harvesting infrastructures. Some of the existing water harvesting infrastructures have been managed by water user groups and committees. However, weak or absent regulatory instruments and human capacity makes management of water harvesting infrastructure a challenge in several intervention villages. Poor enforcement of bylaws also contributes to poor maintenance of the water harvesting infrastructure.

Conclusion

The extent of adaptation and adoption of external catchment water harvesting technologies in Tanzania has been studied 10 years after the end of an extended period of active intervention, with the main focus on floodwater diversion systems and micro-dams (*ndivas*). Most of the interventions that have been done on water harvesting technologies at field and catchment scales were through external interventions such as research projects and NGOs. At national level, the emphasis of government agencies continues to be on developing irrigated

Table 8.3 Possible improvements to water harvesting systems in the Makanya catchment

Technology	Possible improvements
Micro-dams (*ndivas*)	• Repair intake structures • Construct division boxes for secondary canals • Construct lined canals from the sources to the *ndivas* • Construct lined canals downstream of *ndivas* to distribution chambers • Construct water distribution gates and flow measuring devices • Construct silt traps to reduce maintenance effort • Irrigation scheduling based on crop water requirement • Advice on possible high value crops • Proper land preparations improve water distribution • Soil conservation measures (e.g. mulching, terracing)
Floodwater diversion	• Gully improvements using gabion walls or stone pitching to stabilize gully banks and divert water into fields and minimize bank erosion • Repair existing intake structures • Construct robust intake structure: construct gate mechanism such that the gate can be lifted to allow for the siltation to be flushed away by flood • Design and installation of detachable canal gates that can be removed and stored during off-season to avoid vandalism • Development of village water plan to locate areas for farmers, pastoralists, residential use and industry • Reinforcement of banks of the main canals using gabions or stone pitching on high impact areas (e.g. on curves) • Construction of Chaco Dams for livestock to avoid damage of irrigation infrastructure when livestock are searching for drinking water

farming rather than investing in water harvesting technologies. The sustainability of water harvesting interventions is not assured simply because of the apparently simple technology and requires a participatory approach to design and implementation; farmers' contribution to create a sense of ownership, introduction of service charges to assist in infrastructure rehabilitation; and formation of user groups that will assist in maintaining and developing the infrastructure.

While the main focus of the original research (Gowing et al., 2003; Hatibu et al., 2003) and the recent revisit reported here was on the water harvesting systems introduced to WPLL, there are lessons to be learned from comparison with the *majaruba* system elsewhere in Tanzania, which is a continuing success story. Despite receiving no external support, this system has demonstrated sustainability over more than 50 years. There is no reliable data on which to assess its current extent, but evidence suggests that it continues to spread. Its adoption and spread without external intervention is quite remarkable and can be seen as indicative of the potential of water harvesting practices in semi-arid Tanzania.

Why is the *majaruba* system a success story when efforts to promote adoption of water harvesting systems in WPLL have been more problematic and spread has been slow? Amongst other possible explanations, it is important to recognize essential differences in the technology. For its efficient functioning at landscape scale, the *majaruba* system requires collective land-use planning, but the technology itself is divisible. It is possible for an individual farmer to construct and manage this water harvesting system by diverting local hillslope runoff. Floodwater diversion water harvesting systems (both with and without storage reservoirs) as introduced to WPLL are more complex. These systems divert concentrated runoff flows that are larger and potentially much more damaging than the flows intercepted by *majaruba* systems. Intake structures therefore require design and construction skills beyond what can be expected from farmers; external technical assistance is required. The technology is not divisible and therefore requires collective action amongst a group of farmers for construction, operation and maintenance.

Water harvesting systems of the floodwater diversion type have many of the characteristics of formal irrigation systems in that they are socio-technical constructs requiring both technical and social organization for their success. For this reason, is unlikely that they will spread spontaneously; they require external intervention and therefore compete directly with formal irrigation schemes for support. A recent review of irrigation policy (Therkildsen, 2011) suggests that ideological notions about 'modernizing agriculture' motivated a policy shift towards a push for irrigation after 2005. Land under irrigation increased by 15,000 to 20,000 ha per year from a total of 264,000 ha in 2006 to 332,000 ha in 2010. No similar push for water harvesting was evident despite the earlier policy decision.

Notes

[1] This section has been largely compiled from FAO (2005).
[2] These categories differ slightly from those set out in Chapter 2, Table 2.2. Thus, 'low ratio' covers some 'microcatchment' systems; 'intermediate ratio' embraces some 'microcatchment' systems and some 'external catchment' systems; 'high ratio' covers some 'external catchment' systems and some 'floodwater harvesting' techniques.

References

Bakari, A. H., Mahoo, H. F. and Hatibu, N. (1998) 'Performance of maize under gully flow supplementary irrigation', *Proceedings of Tanzania Society of Agricultural Engineers*, vol. 8, pp. 20–42.

Chiza, C. K. (2005) 'The role of irrigation in agriculture, food security and poverty reduction', Proceedings of the 3rd annual engineers day, vision 2025: engineering contribution in poverty reduction, Karimjee Hall, Dar es Salaam, Tanzania.

Court, J. and Young, J. (2003) 'Bridging Research and Policy: Insights from 50 Case Studies', ODI, Working Paper 213, London, UK.

Critchley, W., Reij, C. and Seznec, A. (1992) 'Water harvesting for plant production, vol. II: Case studies and conclusions from Sub-Saharan Africa', World Bank Technical Paper No. 157, World Bank, Washington D.C.

Enfors, E. I. and Gordon, L. J. (2008) 'Dealing with drought: The challenge of using water system technologies to break dryland poverty traps', *Journal of Global Environmental Change*, vol. 18, pp. 607–616.

FAO (2005) *Irrigation in Africa in figures*, AQUASTAT Survey – 2005. FAO Water Report 29, Tanzania, pp. 597–608, available at: http://www.fao.org/nr/water/aquastat/countries_regions/tanzania/index.stm

Gowing, J. W., Mahoo, H. F. and Mzirai, O. B. (1999) 'Review of rainwater harvesting techniques and evidence for their use in semi-arid Tanzania', *Tanzania Journal of Agricultural Science*, vol. 2, no. 2, pp. 171–180.

Gowing, J. W., Young, M. D. B., Rwehumbiza, F. B., Mzirai, O. B. and Hatibu, N. (2003) 'Developing improved dryland cropping systems in semi-arid Tanzania: use of a model to extrapolate experimental results', *Experimental Agriculture*, vol. 39, no. 3, pp. 293–306.

Hatibu, N., Young, M. D. B., Mahoo, H., Gowing, J. W. and Mzirai, O. B. (2003) 'Developing improved dryland cropping systems in semi-arid Tanzania: experimental evidence for the benefits of rainwater harvesting', *Experimental Agriculture*, vol. 39, no. 3, pp. 279–292.

IFAD (2007) *United Republic of Tanzania: Participatory Irrigation Development Project (PIDP)*, Report no. 1831-TZ – Rev. 1, May 2007, IFAD, Rome, Italy.

Kanyeka, Z. L., Msomba, S. W., Kihupi, A. N. and Penza, M. S. (1994) 'Rice ecosystems in Tanzania, characterisation and classification', *Tanzania Agricultural Research and Training Newsletter*, vol. 9, pp. 13–15.

Kato, M. (2001) 'Intensive cultivation and environment use among the Matengo in Tanzania', *African Study Monographs*, vol. 22, no. 2, pp. 73–79.

Kauzeni, A. S., Kikula, I. S. and Shishira, E. K. (1987) 'Development in soil conservation in Tanzania', in: P. Blaikie (ed) *History of Soil Conservation in the SADCC Region*, SADCC Soil and Water Conservation and Land Utilisation Programme, Report No. 8, Maseru, Lesotho.

Komakech, H., van Koppen, B., Mahoo, H. and van der Zaag, P. (2010) 'Pangani River Basin over time and space: On the interface of local and basin level responses', *Agricultural Water Management*, vol. 98, no. 11, pp. 1740–1751.

Lazaro, E. A., Senkondo, E. M. M., Bakari, A., Kishebuka, S. R. and Kajiru, G. J. (1999) 'A small push goes a long way: farmers' participation in rainwater harvesting technology development', *Tanzania Journal of Agricultural Sciences*, vol. 2, no. 2, pp. 219–226.

Lutkamu, M. H., Shetto, M. C., Mahoo, H. F. and Hatibu, N. (2005) 'Scaling-up and uptake promotion of research findings on NRM in Tanzania', East Africa Integrated River Basin Management Conference, SWMRG-SUA, Nairobi, Kenya.

Makurira, H., Mul, M. L., Vyagusa, N. F., Uhlenbrook, S. and Savenije, H. H. G. (2007) 'Evaluation of community-driven smallholder irrigation in dryland South Pare Mountains, Tanzania: A case study of Manoo microdam', *Physics and Chemistry of the Earth*, vol. 32, pp. 1090–1097.

Makurira, H., Savenije, H. H. G., Uhlenbrook, S., Rockström, J. and Senzanje, A. (2009) 'Investigating the water balance of on-farm techniques for improved crop productivity in rainfed systems: A case study of Makanya catchment, Tanzania', *Physics and Chemistry of the Earth*, vol. 34, pp. 93–98.

Mbilinyi, B. P., Tumbo, S. D., Mahoo, H. F., Senkondo, E. M. and Hatibu, N. (2005) 'Indigenous knowledge as decision support tool in rainwater harvesting', *Physics and Chemistry of the Earth*, vol. 30, no. 11–16, pp. 792–798.

Meertens, H. C. C., Ndege, L. J. and Lupeja, P. M. (1999) 'The cultivation of rainfed lowland rice in Sukumaland, Tanzania', *Agriculture, Ecosystems and Environment*, vol. 76, pp. 31–45.

Mutiro, J., Makurira, H., Senzanje, A. and Mul, M. L. (2006) 'Water productivity analysis for smallholder rainfed systems: a case study of Makanya catchment, Tanzania', *Physics and Chemistry of the Earth*, vol. 31, pp. 901–909.

Pachpute, J. S., Tumbo, S. D., Sally, H. and Mul, M. L. (2009) 'Sustainability of rainwater harvesting systems in rural catchment of Sub-Saharan Africa', *Water Resources Management*, vol. 23, no. 13, pp. 2815–2839.

PIDP (2007) 'Lessons Learnt from Development Projects in Agricultural Water Management: A case study of Smallholder Flood Plains Development Programme (SFPDP) of Malawi and Participatory Irrigation Development Programme (PIDP) of Tanzania', Expert Consultative Meetings held in Morogoro, Tanzania 9–10 July and Lilongwe, Malawi 12–13 July 2007.

Shaka, J. M., Ngailo, J. A. and Wickama, J. M. (1996) 'How rice cultivation became an "indigenous" farming practice in Maswa district, Tanzania', in: C. Reij, I. Scoones, and C. Toulmin (eds), *Sustaining the Soil, Indigenous Soil and Water Conservation in Africa*, Earthscan, London.

Stenhouse, A. S. (1944) 'Agriculture in the Matengo Highlands', *East Africa Journal*, vol. 10, pp. 22–24.

SWMRG (2001) *Assessment Analysis of RWH Demand and Efficacy*, Final Technical Report to DFID, Sokoine University of Agriculture (SUA), Soil-water Management Research Programme, November 2001.

SWMRG (2005a) *Improvement of Soil Fertility Management Practices in Rain Water Harvesting Systems: Annex B1-B10*, Final Technical Report, NRSP Project R8115, Sokoine University of Agriculture (SUA), Soil-water Management Research Programme, August 2005.

SWMRG (2005b) *Improving Management of Common Pool Resources in Rainwater Harvesting Systems*, Final Technical Report, Natural Resources Systems Programme R8116', Sokoine University of Agriculture (SUA), Soil-water Management Research Group (SWMRG), October, 2005.

Temu, A. E. M. and Bisanda, S. (2006) 'Pit cultivation in the Matengo highlands of Tanzania', in C. Reij, I. Scoones, C. Toulmin (eds), *Sustaining the Soil, Indigenous Soil and Water Conservation in Africa*, Earthscan, London.

Therkildsen, O. (2011) 'Policy making and implementation in agriculture: Tanzania's push for irrigated rice', DIIS Working Paper 26, Danish Institute for International Studies, Copenhagen.

Thornton, R. J. (1980) *Space, Time and Culture Among the Iragw of Tanzania*, Academic Press, New York.

Tumbo, S. D., Mutabazi, K. D., Byakugila, M. M. and Mahoo, H. F. M. (2010) 'An empirical framework for scaling-out of water system innovations: Lessons from diffusion of water system innovations in the Makanya catchment in Northern Tanzania', *Agricultural Water Management*, vol. 98, no. 11, pp. 1761–1773.

URT (2002) *The National Irrigation Master Plan in the United Republic of Tanzania*, Ministry of Agriculture and Food Security (MAFS), Tanzania.

Willcocks, T. J. and Gichuki, F. (eds) (1996) 'Conserve water to save oil and the environment', Report IDG/96/15, Silsoe Research Institute, UK.

Chapter 9

Sudan

Ancient traditions receiving a new impetus

William Critchley, Abdelaziz Gaiballa,
Abdalla Osman Eissa, Asha Mohamed Deen
and Eefke Mollee

Introduction

The Sub-Saharan Water Harvesting Study (SSWHS) considered that 'Sudan has probably the most extensive and diverse heritage of traditional water harvesting and water spreading of any country in Sub-Saharan Africa' (Critchley et al., 1992). Without the benefit of precise data, this almost certainly remains uncontested for the 'new' Sudan, after the 2011 sub-division of the former country into Sudan and South Sudan. Apart from large-scale irrigation from the River Nile

Figure 9.1 Country map of Sudan

and some relatively minor irrigation from pumped groundwater, Sudan's only other means of growing crops is through careful *in situ* rainfall conservation; or in the very extensive semi-arid and arid zones, a variety of water-harvesting techniques – there is no other option. Under the SSWHS of the later 1980s, reported in Critchley et al. (1992), the two most important water harvesting technologies in Sudan, namely the *teras* system and floodwater harvesting were investigated. This chapter reflects on these two systems – in the context of the new Sudan some 25 years later, and in the light of a growing population, increasing demands on resources and the risks associated with climate change.

Sudan: background[1]

Sudan has a sub-continental climate. Annual rainfall varies from about 800 mm in the south of the country to 25 mm northwards to the border with Egypt. The rainy season length is limited to three to four months, depending on the part of the country, with the rest of the year virtually dry. However rainfall usually occurs in isolated showers, which vary considerably in duration, location and from year to year. Potential annual evapotranspiration ranges from 3000 mm in the north to 1700 mm in the extreme south of what is now South Sudan. Most agricultural activities are concentrated in the semi-arid and savannah zone, through which the Blue Nile, White Nile and the Atbara rivers flow. The growing season in the region is around four months. The major limiting factor is not the agricultural potential of the soils, but the short duration of the rainy season and the erratic distribution of rainfall during the growing period.

The population of the country after separation of South Sudan, was about 33.4 million. Most of the population lives along the Nile and its tributaries, while some live around water points scattered around the country. Poverty in Sudan is predominantly a rural phenomenon. Cultivation is mainly for household subsistence needs. In 2002, before the sub-division of the country, the former Sudan's cultivated land was estimated at about 16.65 million hectares (ha) (seven per cent of the total land area and 16 per cent of the cultivable area).

Rainfed agriculture covered the largest proportion of cultivated land in the former Sudan: it almost certainly still does in the new Sudan. The area actually cultivated and total yield varies considerably from year to year depending on, generally, rainfall. The rainfed farming systems are characterized by small farm size, labour-intensive techniques employing hand tools, low input levels and poor yields. Crops grown in the rainfed sector include sorghum, millet, sesame, sunflower and groundnuts. According to the latest estimates for the former Sudan, the traditional rainfed farming sector contributed all the production of millet, 11 per cent of sorghum, 48 per cent of groundnuts and 28 per cent of sesame production. Mechanized rainfed agriculture comprised about 10,000 large farmers with farm sizes of 400 to 850 ha and a few large companies with holdings of 8400 to 84,000 ha.

Sudan – as it is today – has the largest irrigated area in Sub-Saharan Africa and the second largest in the whole of Africa, after Egypt; this was equally true of the former Sudan. The irrigated sub-sector plays a very important role in the country's agricultural production. Although the irrigated area constituted only about 11 per cent of the total cultivated land in the former Sudan, it contributed more than half of the total volume of the agricultural production. Irrigated agriculture has become more and more important over the past few decades as a result of drought and rainfall variability and uncertainty. It remains the central option to boost the economy in general and increase the living standard of the majority of the population.

Sudan is generally self-sufficient in basic foods, albeit with important inter-annual and geographical variations, and with wide regional and household disparities in food security prevailing across the country. The high-risk areas are North Kordofan, North Darfur, the Red Sea, Butana and the fringes of the major irrigation schemes. Major constraints to higher farm productivity and incomes are high marketing margins on agricultural produce and an inadequate allocation of budgetary resources and of the scarce foreign exchange earnings. As a result, the low input/low-productivity model of production continues to prevail, and small farmers' incomes remain depressed. In the wake of the food shortages experienced in the 1980s, high priority has been given by the government to producing food crops. This has resulted in large expansions in sorghum and wheat areas and output. Much of this has been at the expense of the main cash crop, cotton, with production declining by more than 40 per cent since the mid-1980s.

Internally generated water resources in Sudan are very limited. The erratic nature of the rainfall and its concentration in a short period places Sudan in a vulnerable situation, especially in rainfed areas. Surface water in Sudan comprises the Nile river system and other, non-nilotic water courses – especially the Gash and the Baraka in the east of the country. These are characterized by large variations in annual flow, and heavy silt loads. The deltas of both, the Gash and the Tokar respectively, are renowned for their spate irrigation systems (see following section on floodwater harvesting).[2]

In the post-colonial period, it was assumed that the only sure way to bring about development would continue to be through large irrigation development. In the former Sudan large-scale irrigated agriculture expanded from 1.17 million ha in 1956 to more than 1.68 million ha by 1977. The 1980s were a period of rehabilitation, with efforts to improve the performance of the irrigation sub-sector. In the 1990s, some smaller schemes were licensed to the private sector, while the four big schemes of Gezira and Managil, New Halfa and Rahad remained under government control because they were considered strategic schemes. In 2000, the total area equipped for irrigation was 1.86 million ha, comprising 1,730,970 ha equipped for full or partial control irrigation and 132,030 ha of spate irrigation: note that this latter figure includes at least some of the 'floodwater harvesting' discussed in the next section. These data remain basically accurate for the current Sudan where all the large-scale irrigation is located.

The main irrigated crops are sorghum, cotton, fodder, wheat, groundnuts and vegetables. Other crops grown under irrigation are sugar cane, maize, sunflower, potatoes, roots and tubers and rice. Irrigated agriculture has been Sudan's largest economic investment, yet returns have been well below potential. The overall objective of water management policies is to improve water use efficiency in agriculture, which includes efficient control of water in the irrigation networks, maintenance of the irrigation structures, provision of technical capacities capable to operate the systems, and efficient and economical maintenance of the irrigation system. Water harvesting has never received high priority nationally. However, in the following section its local importance will become apparent.

History and importance of water harvesting practices

The Teras[3] system

On the plains east of the Nile, and especially in the Butana area, and to the north-east of Kassala in the Border Region, the predominant form of small-scale rainfed farming is supported by water harvesting under the traditional *teras* system (Randall, 1963). Mohamed (1996) confirms that the system is also indigenous, and widespread too, in Central Darfur. It is a form of external catchment water harvesting well suited to these plains where considerable runoff is generated despite slopes of one percent or less.

The traditional *teras* system impounds fields with three-sided earth bunds, capturing surface runoff from an external catchment. Plots – comprising a single *teras* – are usually around one hectare or a little more in size. The bottom bund, sited approximately along the contour, is longer than the upslope 'arms' which are set at right angles, embracing overland flow from the upslope external catchment: the catchment to cultivated area ratio is in the order of 2:1. In the case of a *teras* filling up with runoff/rainfall, the excess water finds its way out around the tips of the arms. The earth bunds are between 0.35 and 0.50 cm in height, and the base width from 1.5 to 2.0 m. Traditionally these were constructed by hand, though mechanization (disc ploughs and even front-loader tractors) had been introduced at the time of the SSWHS investigation in 1988. It is important to note a non-technical issue also: management of the plots is on an individual, or family, basis.

Probably the first mention of the system in the English language was by Makinnon in 1948 who, describing how farmers manage to cultivate under the low rainfall conditions of Butana Region, writes: 'Cultivation is ... mainly confined to flooded *wadis* or to areas where *terūs* can be constructed to conserve the rainfall.' Presumably Makinnon – author of the chapter on Kassala Province in Tothill's *Handbook of Agriculture in The Sudan* – also supplied his editor with the entry for the book's glossary. Here we find: '*terās*, pl. *terūs,* col. Ar., a small earth bund built with hand tools for impounding rain-water for watering rain-grown *dura*' (Tothill, 1948).[4]

Barbour (1961) also briefly covers the system in his geography textbook, and includes a photograph of harvested *terūs* in Central Gezira. However, Randall (1963) was the first to describe in detail, and illustrate, the indigenous *teras* system. A senior lecturer in geography based at the University of Khartoum, Randall took the opportunity to present his investigations at the ninth annual conference of the Philosophical Society of The Sudan held in 1961, the proceedings of which were edited and published by Lebon in 1963. He introduces the *teras* system by describing how this technique of water harvesting through creating 'simple three-sided fields' underpinned sedentary agriculture in Blue Nile and Kassala Provinces: 'The expansion of permanent settlement was dependent upon the artificial concentration of runoff upon a given area, an end achieved by impounding the water behind earthen ridges 12–18 inches high, thrown up, in theory at least, along the contour' (Randall, 1963).

Randall (1963) went on to explain the various possible configurations of a single *teras*. The *sadra* is the term for the catchment area outside the cultivated plot, and the *hugna* is the zone where the runoff water concentrates within. In certain situations, in order to spread the benefits better in a good year, 'mini-*teras*' termed *gata'a* are constructed within the main *teras*. He illustrated seven different configurations: specific design being a function of topography, rainfall, slope and other factors including, no doubt, individual preference. The main crop produced, he noted, was short-duration sorghum – with several varieties mentioned. Cotton was apparently also grown in *teras*. And finally, okra was planted around the *teras* bunds where this much-favoured vegetable could gain access to sufficient water.

It was not until a quarter of a century later that attention was brought more widely to the importance of this system. Pacey and Cullis (1986) – who did much to renew interest in water harvesting in Africa through their pivotal publication – quote Wickens and White (1978) with respect to the *teras* technique. Similarly to Randall (1963), Wickens and White describe embankments on plains to the east of the Nile that 'intercept sheet-wash runoff following heavy storms'. The Sub-Saharan Water Harvesting Study then selected the system as one of 13 case studies in sub-Saharan Africa (Critchley et al., 1992). While pointing out that the *teras* system is 'one of the few examples of traditional water harvesting spread over a very extensive area' they emphasised that not enough was known about extent, dynamics, as well as benefits and costs. Despite the important research conducted and reported in van Dijk and Ahmed (1993) the same can be reiterated today.

Van Dijk and Ahmed (1993) consider the systems to be several centuries old, and to have evolved from rectangular *in situ* water conservation basins into fields with the upper side open to accept runoff from a catchment. Though the agropastoral Beja tribes of the Hedendoa and the Beni Amer predominate in this area and make the most use of this production system, the authors note that contributions to the design may have been made by West African pilgrims who passed through the area on their way to Mecca; a number settling around Kassala on their way back, rather than returning home. Van Dijk and Ahmed (1993) emphasize the

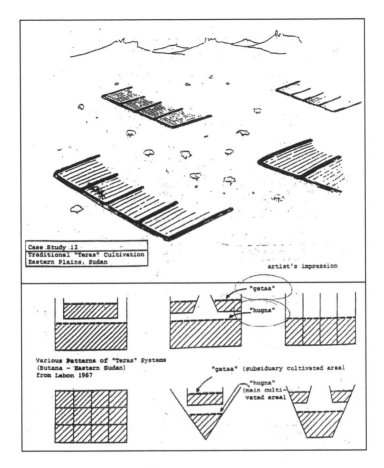

Figure 9.2 Teras (Critchley et al., 1992)

extent of *teras* in the Border area – east of the Gash river which covers some 8600 km² and has an average annual rainfall (over the period from 1938 to 1990) of 286 mm. Production figures are given as between 250 and 500 kg/ha of sorghum (it should be noted that the catchment area is excluded from this calculation). An important by-product is sorghum stover which is fed to live-stock. It is also worth noting that a 'failed' crop can also yield a significant amount of fodder.

The average size of *teras* in the Border area is 1.6 ha of cultivated land which is supplied by an external catchment of 4.9 ha. Further data given by van Dijk and Ahmed (1993) concern the labour requirements to construct *teras*. Initial invest-ment when built by hand incurs 36–95 hours of work per hectare. Seasonal main-tenance depends on the degree of damage caused by runoff: 'If necessary, bunds are raised, arms remodeled and catchments cleared of vegetation.' In contrast to floodwater harvesting schemes, *teras* construction and repair is the responsibility of the individual farmer and is relatively low cost and straightforward.

Floodwater harvesting

Pacey and Cullis (1986) briefly touch on the technique of floodwater harvesting – achieved through the construction of dams or barrages to divert water from ephemeral water courses onto cultivated land. They highlight examples from the Negev desert of Israel and South Yemen rather than Africa. However their observation that 'many other countries' also utilize such techniques is clearly applicable to Sudan, as it is to Ethiopia, Eritrea and Somalia. Allan and Smith (1948) write of 'flush irrigation' and describe it as being 'essentially a development from natural flooding … being merely a single soaking of the land'. In the west of Sudan, '*wadi* agriculture' is reported, and according to both Widanapathirana (1985, 1986 and 1987), and Quin and Willcocks (1989) there is much unexploited potential. Critchley et al. (1994) illustrate three variations on *wadi* agriculture based on Widanapathirana (1987), and point to the importance of floodwater spreading from *khors* in eastern Sudan and notably the Red Sea Hills. Soghayroun (2010) has recently described a variety of floodwater harvesting methods, in the context of the historical importance of *wadis* with respect to settlement and trade in Sudan.

Around Kassala, in the 1980s, there were government attempts to introduce floodwater harvesting schemes. At the end of that decade a pilot research scheme was established after approval by the National Council for Research and the Ford Foundation. The 'Water Spreading Research Kassala' (WARK) as it was known, started in 1989 and was active for two seasons. However this attempt, at Hedadeib, to construct diversions dams and low broad-based spreader embankments was unsuccessful due to multiple breaches in the bunds. The main problems cited were unsuitable soils for construction, and inadequate design: the evaluation of the process called for a focus on *teras* in this area, rather than on floodwater spreading (van Dijk, 1991). A crucial point in design of floodwater spreading is to slow the water enough so that it deposits fine materials – and doesn't cause erosion. This might not have been achieved at this site (Sayed Dabloub, personal communication).

Van Steenbergen et al. (2010) in their comprehensive guidelines to spate irrigation tabulate the 'area equipped for spate irrigation in selected countries' with Sudan falling just behind Somalia in extent. These two countries lead the African statistics, though fall short of Yemen and a long way behind Pakistan. The principle source of this data, the land and water database of FAO, namely FAO (2005), attributes the (strangely precise) figure of 132,030 ha to Sudan (see section 2). As van Steenbergen and colleagues (2010) point out, there is considerable uncertainty about the aerial coverage of such schemes, worldwide. It may be self-evident but is worth pointing out that the statistics available 'do not always capture the smaller, farmer-managed, informal schemes'. The extent of informal spate irrigation/floodwater harvesting is surely far more than the figures cited for

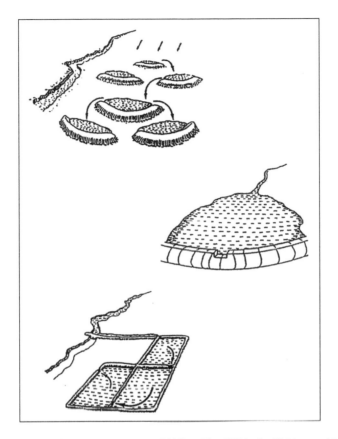

Figure 9.3 Floodwater harvesting variations (W. Critchley, 1994; after Widanapathirana, 1987)

Sudan, and its potential for expansion and improvement undoubtedly much greater.

A series of pertinent points about spate irrigation are made by van Steenbergen et al. (2010). Three are selected here for their direct relevance to Sudan. The first is non-technical: in contrast to the *teras* described above – within spate irrigation, an understanding of the socio-economic context in which farmers jointly operate systems is essential. The second concerns soils, and has been alluded to already in this chapter. In many spate areas they write, fertility is generally not an issue: '[F]ertility is ensured by the regular replenishment of fine silts, carrying organic material eroded from the catchments.' They echo the words of Allan and Smith (1948) regarding what used to be termed 'flush irrigation' where 'the method has the merit[s] of ... covering the land with a deposit of fertilizing silt'. Third, they point out that there is a large potential for other contributions to improved spate irrigation, including agronomy, post-harvest technology and agroforestry. This is the 'WH+' referred to in Chapter 2.

Key water-harvesting developments: the technologies and the approaches

The Teras system[5]

Teras continue to be the mainstay of non-irrigated agriculture in Kassala State, which was reinvestigated 25 years after the SSWHS study. Van Dijk and Ahmed (1993) had studied aerial photographs of the Border Region and these showed little change in the area of *teras* between 1963 and 1986. In the early 1990s these authors found calculated the total area under *teras* in this region to be an estimated 20,000 to 40,000 ha covering two to five per cent of the overall area. They tend, apparently, to expand in number during and (presumably) immediately after years of good rainfall when agricultural investment activity is stimulated by healthy yields. According to the State Ministry of Agriculture and Animal Resources they have increased over the past 20 years and especially recently: figures provided by the Land Use and Desertification Control Branch (LUDC) indicate that approximately 10,000 ha of *teras* have been constructed for impoverished farmers since 2007 under various agencies: both international and national.

There is a branch of the Eastern Region Development Programme (ERDP), funded by the EU, in Kassala, and it is involved in building both *teras* but some water spreading schemes also for the target group of 'poor rural families and female headed households' (Euroconsult Mott Macdonald, 2011). Hand tools and seeds are also supplied. In 2011, 400 ha of *teras* (included in the 10,000 ha cited above) were constructed through the LUDC under the ERDP. An example of this implementation can be found at Tajouj, close to the Eritrean border. Visiting the area during the first cropping cycle, during a period of drought stress, gave a very clear visual impression of the impact of the harvested overland flow (personal observation: Critchley and Gaiballa). At the lower end (the bottom bund is constructed more or less on the contour) of the three-sided cultivated field, where most runoff had been captured, the sorghum crop was taller and stronger. Indeed there is often a very distinct continuous gradation from the top of the impounded area (with poor crops) to the bottom (with good crops). Such a transect testifies to the crop response to harvested runoff under water-stress. Yields records for the 2011 season, which was below average in rainfall, are quoted as 1250 kg/ha of sorghum (Euroconsult Mott Macdonald, 2011).

There are three key recent developments. First, a technical improvement: a new tool – a blade front-mounted on a tractor – has been developed to build the bunds. This is an advance from the front-loader seen in operation some 25 years previously on the occasion of the SSWHS visit. Second, there are several NGOs operational in Kassala, but all of these are coordinated through the Ministry of Agriculture and Animal Resources as far as water harvesting activities are concerned. No longer are NGOs permitted to 'go it alone'. This is a development that has consequences that stretch farther than water harvesting; it is illustrative of a general tendency in Sub-Saharan Africa to 'harness' NGO-activity rather than

Photo 9.1 Farmer with sorghum grown in Teras near Kassala (W. Critchley)

to let NGOs go their own way. There are associated important implications for norms and standards, both in terms of technologies and approaches. Thirdly, there is an apparent willingness on behalf of the government to construct *teras* structures for farmers who lie on the outer margins of sustainable livelihoods; those unable to fend for themselves during periods of instability and associated poverty. Naturally this refuels the age-old development debate regarding the sustainability of investments made through aid, without significant local contribution. In theory this also opens up the possibility of more careful and consistent monitoring.

Floodwater harvesting

Floodwater harvesting from *khors* or *wadis* (ephemeral water courses) is growing in importance.[6] An example of a scheme which was constructed in 1999 is located at Hashitribab, some 7 km from Sirkat on the road towards Kassala. Hashitribab is typical of floodwater harvesting in Red Sea State (here, the annual rainfall is in the order of 120 mm per annum). In this case, as in many others, there used to be a traditional hand-built earthen diversion barrier which had to be reconstructed regularly. The new scheme seeks to improve the intake infrastructure on the basis of technical design considerations and make the barrier more permanent.

This scheme comprises a stone-pitched earth barrier, some 175 m in length with an integral spillway, across a *khor*. The barrier deflects water and leads it along a diversion canal towards a series of 'energy dissipating' stones set in the earth. From this point (after another 200 m) it begins to spread over even terrain until it reaches fields which begin about a kilometre downstream. This water-spreading scheme provides spate flow to fields which extend to some 200 ha, benefiting some 300 farm families. The total cost is estimated at around US$80,000, representing some US$400 per hectare of cultivable land.

While the diversion is still intact (and functional: there were young sorghum plants growing at the time of the visit in November 2011) the need for mainte-nance is evident. The modest input of voluntary labour in its original construction noted above, comprising a contribution to manual stone-pitching of the main bund, constituted a total of (approximately) 7 per cent of overall establishment costs. Construction by the government, using machinery, with little local contri-bution might explain why voluntary maintenance has been negligible. This echoes the free-aid debate – already raised in this chapter – in relation to the construction of *teras* systems for farmers.

Water is held, and spread, in the cultivated areas through various types and configurations of bunds. In the case of Hashitribab, earth bunds, roughly along the contour, serve the purpose. Excess runoff flows around the tips of the bunds and successively reaches fields downstream. Another, more carefully constructed distribution system can be seen close to Port Sudan downstream of a water spreading diversion. In this location, each field has an inlet and an outlet (simple spillways that can be opened and closed). These fields are inun-dated each season with the spate waters and residual moisture serves as the basis to grow sorghum and tomatoes. One important constraint is the accumulation of fine clays within the fields, which can lead to poor infiltration after a number of years. This necessitates some tillage to break the surface pan that forms. Another concern is the potential negative impact on downstream ecosystems – deprived of the water they used to receive. The ERDP is active in Red Sea State, and has a water harvesting component under which these fields had been developed/ rehabilitated, implemented by the Soil Conservation, Land Use and Water Programming Administration (SCLUWPA) under the Ministry of Agriculture and Animal Resources.

Water spreading schemes have gradually expanded in number over the last 25 years (El Sammani and Dabloub, 1996; Sayed Dabloub, personal communica-tion). Significantly there has been a recent upsurge in construction activity. It was confirmed that there are various new sites under planning and construction currently – and the driving force currently is the ERDP. Implementation is under SCLUWPA and there are plans to improve and expand 20 sites within the state, totaling some 5500 ha and benefiting 4700 farm families in terms of access to flood waters for irrigation (SCLUWPA, 2010). In Kassala State also, the ERDP (through LUDC) has established three floodwater spreading schemes, at Shalalop, Inkelkalaat and Hasyabab, all in 2011, covering some 760 ha (Euroconsult Mott Macdonald, 2011). This increase in area and importance resonates with the obser-vation made by van Steenbergen et al. (2010) that floodwater harvesting/spate irrigation is 'on the upsurge in several countries, for instance in the Horn of Africa'.

Photo 9.2 Water spreading barrier at Hashitribab (W. Critchley)

Conclusion

These two types of water harvesting in Sudan were highlights of the original World Bank commissioned study (Critchley et al., 1992). In both cases they were clearly vital for the local economy, and crucial for food production. One, the *teras* system makes use of surface runoff and is individually managed; the other floodwater harvesting, captures channel flow and supplies water to adjacent fields cultivated by a community. There are, simply, no alternatives to water harvesting in these parts of Sudan. Both systems thrive, and have expanded in coverage. This may be partially explained by rising population – but there is undoubtedly a renaissance of interest and investment in water harvesting. Floodwater harvesting is the only option in the more arid conditions of the Red Sea Hills. Around Kassala, in the Border Region, there is much to support the policy recommendations put forward by van Dijk and Ahmed in 1993 who emphasize the development of small-scale, individual farmer-managed *teras* rather than floodwater harvesting schemes. Nevertheless, recent experience under ERDP suggests that such water spreading schemes can thrive here also: yields in 2011 were only marginally less than those of the close-by *teras*.

Finally, however convincing the case for water harvesting is in Sudan, there continue to be inadequate data to evaluate the systems in more than basic terms. Research is surely vital, but systematic and sustained monitoring and evaluation systems are equally important: they need to be established on a national basis to collect and analyse essential data, namely extent, impact and progress, on water harvesting.

Notes

1   This section has been largely compiled from three sources, principally FAO (2005), but also World Bank (2010) and CIA (2012), and has differentiated as far as is possible between data for the former, and the new, Sudan.
2   'Spate irrigation', as described by these authors, has a broader definition than that attributed to 'floodwater harvesting' in this chapter and in this book (see Chapter 2). Spate irrigation definitions *include* much floodwater harvesting, but area data from AQUASTAT focus on those spate irrigation schemes with developed infrastructure, and those that come close to conventional irrigation where the water supply is more secure with 'significant base flows': well-known spate irrigation systems in Sudan, which fall outside our definition of floodwater harvesting, but are included in the AQUASTAT data, are the Gash and Tokar delta systems.
3   While Tothill (1948) writes of *terās* in the singular and *terūs* in the plural, Randall (1963), and van Dijk and Ahmed (1993) use *teras* and *terus*. Mohamed (1996) refers to *tera* (singular) and *trus* (plural). This chapter simply uses *teras*.
4   *Dura* (derived from the Arabic) was the common term in use within Sudan, at the time, for sorghum (*Sorghum vulgare*).
5   Data collected and entered in WOCAT's Questionnaires on Technologies and Approaches. Main contributors: Abdalla Osman Eissa and Abdelaziz Gaiballa for Floodwater Harvesting, Red Sea State; Asha Mohamed Deen and Abdelaziz Gaiballa for *Teras*, Kassala State.
6   Soghayroun (2010) in his book that examines the connection between trade and *wadis* doesn't distinguish between the terms – though the named larger watercourses are almost invariably prefixed by '*Wadi*' – and gives the correct usage as *wadi* or *khor* (singular) and *widyan* or *khairan* (plural).

References

Allan, W. N. and Smith, R. J. (1948) 'Irrigation in the Sudan', in: J. D. Tothill (ed) *Agriculture in The Sudan: A Handbook of Agriculture as Practised in The Anglo-Egyptian Sudan*, Oxford University Press, London.
Barbour, K. M. (1961) *The Republic of Sudan. A Regional Geography*, University of London Press, London.
CIA (2012) 'The World Factbook', available at: www.cia.gov/library/publications/the-world-factbook/geos/su.html, accessed 24 May 2012.
Critchley, W. R. S., Reij, C. and Willcocks, T. J. (1994) 'Indigenous soil and water conservation: a review of the state of knowledge and prospects for building on traditions', *Land Degradation and Rehabilitation*, vol. 5, pp. 292–314.
Critchley, W., Reij, C. and Seznec, A. (1992) 'Water harvesting for plant production. Volume II: case studies and conclusions for Sub-Saharan Africa'. World Bank Technical Paper Number 157, Africa Technical Department, World Bank.
El Sammani, M. O. and Dabloub, S. M. A. (1996) 'Making the most of local knowledge: water harvesting in the Red Sea Hills of Northern Sudan', in: C. Reij, I. Scoones and C. Toulmin, *Sustaining the Soil: indigenous Soil and Water Conservation in Africa*, Earthscan, London.
Euroconsult Mott MacDonald (2011) *Provision of Technical Assistance for the Implementation and Management of the Eastern Recovery and Development Programme*. Government of Sudan and European Development Fund. Progress Report No. 5: July–December 2011.

FAO (2005) *Irrigation in Africa in figures, AQUASTAT Survey – 2005*. FAO Water Report 29, Sudan, pp. 527–541.

Makinnon, E. (1948) 'Kassala Province.' in: J. D. Tothill (ed.), *Agriculture in The Sudan: A Handbook of Agriculture as Practised in The Anglo-Egyptian Sudan*, Oxford University Press, London.

Ministry of Information, Government of Sudan (2011) *Sudan: the Land of Opportunities*.

Mohamed, Y. A. (1996) 'Drought and the need to change: the expansion of water harvesting in Central Darfur, Sudan' in: C. Reij, I. Scoones and C. Toulmin (eds), *Sustaining the Soil: indigenous Soil and Water Conservation in Africa*, Earthscan, London.

Pacey, A. and Cullis, A. (1986) *Rainwater Harvesting: The Collection of Rainfall and Runoff in Rural Areas*, Intermediate Technology Publications, London.

Quin, M. and Willcocks, T. J. (1989) *Sudan: development of crop production on Naga'a and other alluvial soils in South Darfur*, Overseas Division, AFRC Engineering Report to ODA, OD/89/23, ODA, London.

Randall, J. R. (1963) 'Land Use on the Arid Margin of the Clays in Blue Nile and Kassala Province' in: J. H. G. Lebon (ed.), *Proceeding of the 9th Annual Conference Philosophical Society*, Sudan, 1961.

SCLUWPA (2010) *Preliminary Assessment Report in Preparation for Water Harvesting Activities under the Eastern Recovery Development Programme (ERDP)*, SCLUWPA, Port Sudan.

Soghayroun, I. S. E. (2010) *Trade and Wadis System(s) in Muslim Sudan*, Fountain Publishers, Kampala.

Tothill, J. D. (ed) (1948) *Agriculture in The Sudan: A Handbook of Agriculture as Practised in The Anglo-Egyptian Sudan*, Oxford University Press, London.

van Dijk, J. A. (1991) *Water Spreading by Broad-Based Earth Embankments*, WARK Final Report. National Council for Research, Sudan.

van Dijk, J. A. and Ahmed, M. H. (1993) *Opportunities for Expanding Water Harvesting in Sub-Saharan Africa: The Teras of Kassala*. IIED Gatekeeper Series no. 40, London.

van Steenbergen, F., Lawrence, P. and Haile, A. M. (2010) 'Guidelines on spate irrigation', FAO Irrigation and Drainage Paper 65, Food and Agriculture Organisation of the United Nations, Rome.

Wickens, G. E. and White, L. P. (1978) 'Land-use in the southern margins of the Sahara', in: B.H. Walker, *Management of semi-arid ecosystems*, Elsevier Scientific Publications, Amsterdam.

Widananpathirana, A. (1985) 'Agricultural practices influencing desertification in the Dafur Region, Sudan', Paper presented to the workshop on Sand Transport and Desertification in Arid Lands, November 1985, Khartoum, Sudan.

Widananpathirana, A. (1986) 'Study of present status and proposed development strategy: wadi agriculture in Northern Dafur, Sudan', unpublished paper.

Widananpathirana, A. (1987) 'Exploring wadi resources for irrigation in arid areas of the Western Sudan: strategy and problems', Paper presented at the International Symposium on the Conjunctive Use of Surface and Groundwater', Lahore, Pakistan.

World Bank (2010) *Country data Sudan*, available at: http://data.worldbank.org/country/sudan, accessed 24 May 2012.

Chapter 10

Zimbabwe

Keeping runoff on the land

*Douglas Gumbo, Denyse Snelder, Menas Wuta
and Isaiah Nyagumbo*

Introduction

In their study of 1992, Critchley and his team characterized Zimbabwe as a country with a long history of on-farm soil conservation measures, aimed at soil conservation rather than moisture conservation, located in the wetter areas of the country. Traditional small-scale water harvesting for crop production in drier areas was non-existent, or limited to isolated fields. It was only in the late 1980s that government-initiated projects began experimenting with water-harvesting techniques in these areas. One documented example was the government initiative on 'tied furrows', a small-scale microcatchment water harvesting technique that was tested, both on-station and on-farm, at the Chiredzi research station in the southern semi-arid part of the country. These systems were studied and described by Critchley et al. (1992).

During the last two to three decades, there have been, however, more frequent reports of interventions in the form of water harvesting mainly driven by successive droughts (Mutekwa and Kusangaya, 2006; Kahinda et al., 2007). Whereas in the past the cultivation of drought-tolerant crops was considered as the main solution to intra-seasonal dry spells, harvesting of runoff water was increasingly perceived as a valuable technology to enhance water productivity in drought-prone areas. The 'dead-level contour' (DLC) is an example of a within-field catchment water harvesting system introduced in the late 1980s into some communal areas by NGOs in collaboration with the Department of Agricultural, Technical and Extension Services (AGRITEX) and researchers (Kronen, 1994; Motsi et al., 2004).

However, despite their apparent benefits, water harvesting technologies have not yet been widely adopted in Zimbabwe (Motsi et al., 2004; Mugabe, 2004; Mutekwa and Kusangaya, 2006,) and still form a subject of ongoing research (see Sepaskah and Fooladmand, 2004; Walker et al., 2005; Mupangwa et al., 2006, 2011). Moreover, the political turmoil of the last few years has not created a facilitating environment for technology investment, and has hampered farmer extension and support programmes. There are no resources allocated to developing water harvesting systems that are better suited and more affordable to smallholder farmers, and institutional frameworks have failed to effectively deliver services to these farmers due to lack of human and financial capital (Nyagumbo and Rurinda, 2011).

This chapter discusses the two types of water harvesting systems described above: (1) tied furrows, with a runoff area as tested in Chiredzi District (Masvingo Province); and (2) dead-level contours modified for specific soil types. The chapter will reflect on their status under conditions of growing populations, increasing demands on resources, and climate change. The analysis is based on a review of the relevant literature, multi-stakeholder and expert meetings, on-farm field visits and discussions with farmers during early 2012. The three provinces where the technologies have been introduced are located in the southern part of the country (Figure 10.1), a predominantly semi-arid region characterized (in non-agricultural areas) by *miombo* woodlands and drier savannah vegetation such as *Acacia* spp., *Commiphora* spp. and *Euphorbia* spp.

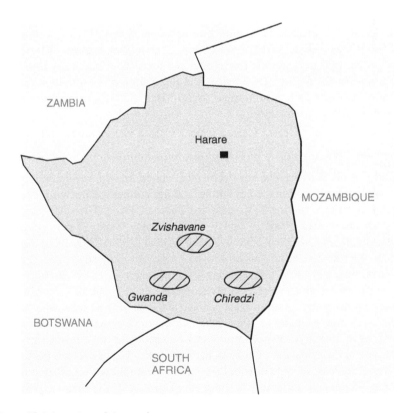

Figure 10.1 Location of the study areas

The selection of the water harvesting technologies in this study was based on the following criteria:

- The technologies have been identified as promising in harnessing water for productive purposes either in the past or in recent times.
- The technologies have been studied before, either in the past (ca. 25 years ago) or in more recent times, and are described in publications that serve as a baseline for the current study.

Following this introduction, the chapter discusses Zimbabwe; its climate, soil environment and farming systems. This is followed by an evolution of policies relevant to water harvesting. A history and importance of water harvesting practices follows, and finally recent key developments are discussed.

Zimbabwe: background

Zimbabwe is a landlocked country, located in southern Africa between a latitude of about 15 and 22° south and a longitude of between 26 and 34° east, with a total area of 386,850 km² (FAOSTAT, 2011). About 30.5 million hectares (ha) (79 per cent) is agricultural land and the remaining area consists of national parks, state forests and urban land. Zimbabwe has a total population of 12.6 million (2010), of which 62 per cent live in rural areas (World Bank, 2012), particularly in the more humid parts of the country.

Climate and soils

The country is divided into five agro-ecological zones, known as Natural Regions (NRs) mainly based on climatic conditions and agricultural productivity declines from NR I (best) to V (poorest) (Vincent and Thomas, 1960). The areas discussed in this study are located in NR IV and V which occupy 38 and 27 per cent of the total agricultural land area respectively (Anderson et al., 1993; FAO, 2006). Natural Region IV with an annual rainfall of 450–650 mm/year is mostly suited to semi-extensive farming and NR V with less than 450 mm/year to extensive farming. Both regions experience high average monthly temperatures of about 38°C, resulting in high potential evapotranspiration rates (up to 1800 mm/year in areas with annual rainfall of 465 mm; Mupangwa et al., 2006). The growing season extends over a period of 90 to 160 days between mid-October and April, but crops suffer from frequent annual droughts and intra-seasonal dry spells. The areas with the lowest rainfall have the least reliable distribution, ranging from 20 per cent variability in the north to 45 per cent variability in the south (Torrance, 1981; Bratton, 1987).

In the drier areas, about 90 per cent of the total rainfall is associated with high intensity storms of short duration. Motsi et al. (2004) note that rainfall intensities greater than 13 mm/h are likely to lead to runoff in all soils, except the sandy soils with high infiltration rates. At least 50 per cent or more of the rainfall has intensities above this level and a high proportion of the rain is lost as runoff. This is exactly where water harvesting can bring benefits: where the runoff can be harvested within or between cropped fields, it can improve soil moisture conditions during intra-seasonal dry spells leading to lower risk of crop failure in semi-arid areas.

Like other countries in the region, Zimbabwe has not been spared from climate change. It is experiencing shifts in the onset of rains, increased frequency of intense rains, more low rainfall years, more tropical cyclones of high intensity, less drizzle, and more frequent and intense mid-season dry spells (Unganai, 1996; Phillips et al., 1998). The IPCC Third Assessment Report suggest that by 2050 temperatures and rainfall over the country will be 2–4°C higher and 10 to 20 per cent less than the 1961–90 baseline, respectively. Agriculture has been identified as the sector most vulnerable to these climatic changes (UNDP, 2010).

Soils in Zimbabwe are variable depending on the geological formation of the area. Granitic rocks are by far the most extensive geological formation and have given rise to light to medium textured soils with high amounts of coarse sand (Nyamapfene, 1991). Most granite-derived soils are characterized by low inherent fertility and low water holding capacities. They generally have deficiencies in nitrogen (N), phosphorus (P) and sulphur (S); and have a low cation exchange capacity (CEC) and water retention capacities, due to low soil organic matter (SOM) and clay contents (Chuma et al., 1997; Nzuma and Murwira, 2000). Where soils are sandy there is also deep percolation, with rainwater quickly disappearing within a few hours of a rainstorm. Basic igneous and metamorphic rock formations are also present, yielding reddish soils with high clay content in well-drained catenal positions.

Agriculture

Just after Zimbabwe gained independence in 1980, there were two dominant and highly contrasting farming sub-sectors in the country. These were: first, the large-scale commercial farm sub-sector (LSCF) consisting of about 4000 farm families as well as around 20 large agro-industrial estates, in total occupying 16.9 million hectares of the total agricultural land. Secondly, there was the 'communal' and the small-scale commercial farming subsector which covered 16.4 million ha hosting a population of 8.4 million people (1.2 million smallholder families). In 1997, the Government of Zimbabwe initiated a process of radical land reform based on extensive compulsory land acquisition and redistribution. The funds for acquiring the land could not be raised and in 2000, the government embarked on the Fast Track Land Resettlement Program (FTLRP). By 2001, two new categories of land use types were created by the government: the A1 and A2 models, in addition to the old commercial and communal sub-sectors. The A1 farmers are essentially smallholder farmers with small individual arable fields and communal grazing areas, held under permit and customary tenure. On the other hand, the A2 sub-sector is made up of self-contained farming units generally about 50 ha or greater.

The farming systems in the communal sub-sector, in the semi-arid areas (NR IV and V) are nowadays still characterized by two major activities: livestock ranching and subsistence cropping. Livestock rearing is the main source of income from agriculture while traditional cropping is targeted mainly at ensuring household food security through the small grains of *Sorghum bicolor* (sorghum), *Pennisetum glaucum* (pearl millet) and *Eleusine coracana* (finger millet). Although *Zea mays* (maize) is not recommended in semi-arid areas, farmers often grow it due to its palatability and high yields in especially good years.

Political turmoil

In the 2000s, the Zimbabwean economy began to deteriorate due to various factors, including mismanagement and corruption, the imposition of sanctions and land reform (Birner and Resnick, 2010). Between 2000 and 2007, the economy shrunk by 50 per cent, inflation rates were among the world's highest and unemployment rates reached 80 per cent in formal jobs (Bird and Busse, 2007; Birner and Resnick, 2010). The country's poor economic condition seriously affected the agricultural sector and associated soil and water conservation programmes (see the next section).

Evolution of policy related to water harvesting

Zimbabwe has a long and politically sensitive history of soil conservation (Dreyer, 1997) stretching from the colonial period, something that it shares with many other countries in Sub-Saharan Africa. During that era, the main conservation efforts were based on the physical protection of arable lands through a system of standard contour ridges linked to waterways[1] also referred to as standard graded contours, to control water disposal and runoff. These systems were derived from those developed in the USA, and their objectives contrasted strongly with the water retention priorities of many farmers in low rainfall areas in southern Zimbabwe (Scoones et al., 1996). However, for decades, the government enforced these essentially 'alien' practices, resulting in the whole concept of soil conservation becoming unpopular among farmers in the communal areas (Elliot, 1987; Wilson, 1988; Hagmann and Murwira, 1996). Contour ridges were introduced indiscriminately for use in smallholder farming areas in the 1930s, and later enforced through the Natural Resources Act section 52 in 1941, without considering the rainfall characteristics that had contributed to accelerated erosion after the introduction of the plough in the 1930s (Aylen, 1941; Alvord, 1958). The use of mechanical contour ridges was resisted by farmers, as it was seen as a tool of oppression and high labour demand, as well as being irrelevant to drought-prone regions. The policies were labelled oppressive, as communal area citizens were not allowed to participate effectively in their formulation and implementation (Mapedza, 2007).

In 1980, when Zimbabwe became an independent state, the government had the authority to formulate new policies for the agricultural sector. Employing a top-down policy formulation technique, with the government secretariat assuming the custodial role (especially regarding the resource-poor farmers), the outcome was a perpetuation of services that benefited the large-scale commercial farmers. Even today, although perceptions and approaches are changing, the standard contour ridges are to some extent still predominant, despite the introduction of other practices in soil and water conservation.

Faced with a potential inability to feed the growing population, the Ministry of Agriculture developed the 1995–2020 agricultural policy framework that mapped the course of agricultural development for the following two decades. The major objectives of this framework were to increase agricultural productivity and ensure food security at household level. Several further policy frameworks on agriculture were announced over a period of 15 years but these have had little impact because they were politically motivated, and enacted merely to justify the chaotic land reform process. The policy frameworks lacked content and specifics on investment and the role of parastatals in increasing productivity of the sector (Birner and Resnick, 2010). During the twentieth century there was technical superiority and higher productivity within the large commercial farming sub-sector, and thus little justification for agricultural policies designed to support smallholder farmers – except to focus on soil conservation measures to protect downstream siltation of major rivers that supplied water to commercial farms and estates. The policy makers in Zimbabwe have consistently used small farm models instead of livelihood framework models and do not recognize rural settlement from a multiple actor perspective (Chimhowu and Hulme, 2006). As a result, some policies aimed at improving agricultural production have contributed to deforestation of woodlands in the communal and resettlement areas of Zimbabwe (Chipika and Kowero, 2000).

Water harvesting: a historical perspective

Water harvesting is referred to as a relatively recent technology in Zimbabwe (Critchley et al., 1992; Mutekwa and Kusangaya 2006), stimulated by recurrent droughts in recent decades. Unlike the Sahelian zones, there is no evidence of traditional systems in water harvesting in the country. From 1959 to 2002, Zimbabwe experienced 15 droughts occurring on average every two to three years (World Bank, 2009). The droughts seriously undermined food security, particularly of those living in the semi-arid regions (Bird and Shepherd, 2003). Rainfed cropping is thus risky in these regions where mid-season dry spells often lead to complete crop failure (Nyagumbo et al., 2009). Hence, farmers have begun to mitigate drought effects by applying water harvesting to increase the time before crop moisture stress starts to set in (Mutekwa and Kusangaya, 2006). Although there are reports of spontaneous uptake (e.g. Mutekwa et al., 2005), most water harvesting technologies have been introduced and promoted sometimes accompanied by subsidies by extension programmes of AGRITEX, under the Ministry of Lands and Agriculture and various NGOs.

In this section both tied furrows and dead-level contours will be discussed, in terms of past (previously described) design, production system and implementation in semi-arid areas of Zimbabwe.

Tied furrow system

In 1988, Critchley and his team examined the tied furrow system in the Chiredzi district where it was introduced through a project of the Department of Research and Specialist Services (DR&SS) and AGRITEX (both under the Ministry of Lands, Agriculture and Rural Resettlement) in 1982. The main features of the system as described in Critchley et al. (1992) are summarized below.

Tied ridging is a well-known system in Africa, but normally the ridges and furrows are narrow and act as *in situ* moisture conservation systems rather than microcatchments. The tied furrow system in Chiredzi was related to tied ridging but differed in that it was wider-based and, hence, functioned as a microcatchment water harvesting technique. It was constructed within fields, with the ridges acting as the catchment and the furrows as the concentration and planting area where rainfall was stored after accumulation and infiltration (see Figure 10.2). The ridges and furrows in Chiredzi were formed by a tractor-drawn ridger and sited on a gradient of approximately 0.33 per cent. The recommended spacing between the ridges was 1.5 meters and the earth ties, which were formed by hand within the furrows, were 15–20 cm high and constructed every 5–10 m, depending on slope (see Figure 10.2). The system was designed to hold a maximum rainfall event of 75 mm (discounting infiltration) before overtopping of the ties allowed safe discharge along the furrows. The effective catchment to cultivated area ratio was at least 1:1.

The seeds (sorghum, maize or cotton) were placed in a line on one side of the furrow approximately 20 cm from the mid-point of the furrow to avoid waterlogging. Plant densities were low and varied from 33,000 or fewer plants per hectare for sorghum and 15,000 or fewer for maize. In the event of adequate rainfall, responsive relay planting (of leguminous crops, typically groundnuts, cowpeas and pigeon peas) was carried out on the opposite side of the furrow. The comparative yields for maize recorded over a four-year period (of two 'dry' and two 'wet' years) were 1415 kg/ha on flat land and 1715 kg/ha on land with furrows.

Farmers' views that the crop spacing was too wide leading to lower yields in a good year (especially of maize) were foreseen as a potential 'stumbling block to easy adoption' by Critchley et al. (1992). During the study in 1988, the system was still limited to fields (approximately eight hectares) of the research station in Chiredzi and 35 on-farm trial plots (of approximately 0.2 ha each). Expansion to other areas and longer-term data gathering were needed to be conclusive about the performance and benefits of the tied furrow system. Yet, at the time of the study, Critchley et al. (1992) noted a considerable demand from the farmers' side for the tractor-drawn ridger to expand the area under tied furrows on their land.

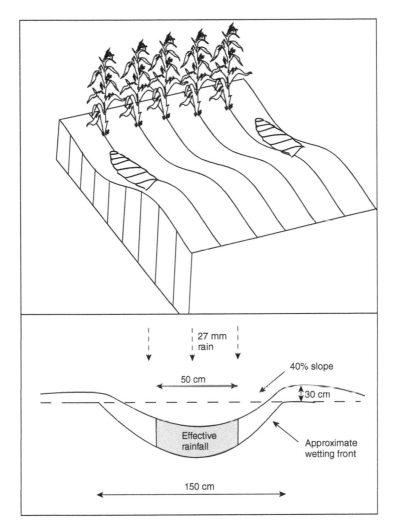

27 mm
rain

40% slope

50 cm

30 cm

Effective
rainfall

Approximate
wetting front

150 cm

Figure 10.2 The tied furrow system: artist's impression (above) and cross-section (below)

However, currently there is no evidence of spontaneous uptake of the technology and the section 'Key water harvesting developments' gives some of the reasons why the technology was largely abandoned.

From standard graded contours to dead-level contours

The technology of dead-level contours (DLCs) is more recent, and is derived from the standard graded contour or contour ridge introduced in the 1930s and enforced at a wider scale since 1941 (see 'Evolution of policy related to water harvesting').

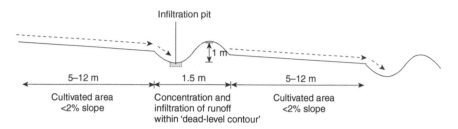

Figure 10.3 Cross-section of a dead-level contour

For understanding the evolutionary process of converting standard graded contours into DLCs, a description of the former is given. A standard graded contour (gradient 1: 250) is an earth ridge made along the contour line within fields. The fields are usually planted with staple food crops (maize and small grains, sorghum and millet). Contour furrows (channels) vary in size, the minimum having a cross-sectional dimension of 1.5 m width and 0.5 m height (Mupangwa et al., 2006) as shown in Figure 10.3. The length of the contour likewise varies depending on field size, and the spacing between graded contours varies with slope gradient and soil type. In the early days of enforced establishment, there was an unwritten rule that the size was adequate if 'a car, driven by the commissioner through the field, could fit in the standard contour channel'. A rule of thumb for the spacing of contour lines is 20–30 m for gentle slopes (gradient: less than five per cent) and 10–15 m for steeper slopes (gradient: more than 10 per cent (Kwashirai, 2006)). The standard graded contour has been classified as a labour-intensive technology, making use of manual labour and also ox-drawn ploughs. In the 1980s and 1990s, however, mechanized construction also took place.

Being primarily designed for soil erosion control and regulated drainage of excess water in wetter areas, the standard graded contour structures proved inappropriate in the drier areas of the country. It is here where in recent times modifications have been introduced to make the structures more effective through strengthening their water harvesting and soil moisture conservation aspects. These innovations, which are described in more detail in the following section, include the construction of a zero-gradient ('dead-level') contour channel that enhances the harvesting and infiltration of water rather than the drainage and removal of water (as in the case of the standard graded contours) received from upslope areas. Thus the system acts as an external catchment system but also captures runoff and soil from within the field. In addition, infiltration pits – either covered (to control evaporation) or uncovered – are dug in the channel at constant distances (minimum 10 metres) depending on the soil texture to further enhance the infiltration and underground storage of water.

So far, little research on dead-level contours and infiltration pits has been conducted in Zimbabwe. However, there are some studies described in recent articles and these focussed on one soil type. For example, the effects of DLCs or infiltration pits on soil moisture retention have reported by Mugabe (2004), Motsi et al. (2004), Mupangwa et al. (2006, 2011). The study of Mugabe (2004) monitored soil water storage above and below infiltration pits in Masvingo communal areas. The results showed that the rain and runoff water captured by the pits slowly seeps downwards to replenish soil moisture lost through evapotranspiration up- and down-slope of the pits. Mupangwa et al. (2011) studied soil water dynamics near DLCs during two growing seasons in the Gwanda district and found that DLCs with infiltration pits captured more rain and runoff water than those without. Significant lateral water movement was observed at 3 m downslope following rainfall events of 60–70 mm/day, with the soil layer between 0.2 and 0.6 m benefiting most. However Mupangwa et al. (2011) conclude that the soil water benefits observed so far, derived from all labour, equipment and time invested in constructing these contours, were short-term, unclear and not worth the investment for smallholder farmers. Given that the study only covered four farms, one soil type and two growing seasons, the findings of Mupangwa et al. (2011) are not conclusive but do suggest caution; they primarily apply to the specific design and conditions under which the contours were studied; that is, semi-arid areas with gentle slopes (< 1 per cent) in light-textured soils and poor marketing and transport systems. Mupangwa et al. (2011) call for more research, exploring new designs of DLCs on different soils and slopes in combination with strip cropping, agroforestry tree species and other *in situ* water management techniques (e.g. mulching, ripping).

The role of socio-economic factors on the uptake of DLCs is described by Munamati and Nyagumbo (2010). Resource ownership was identified as a potential key factor in farmers' ability to upscale water-harvesting techniques since the performance of DLCs was linked to resource status: higher success rates were measured among wealthier households. The women-headed households also proved less successful however; that is, within the most successful group only 5.3 per cent were women because of limited access to resources, especially land.

Key water harvesting developments: the technologies and the approaches

Tied furrow system

While originally the system was found in the districts of Zvishavane and Gwanda and promoted by the Deutsche Gesellschaft für Internationale Zusammenarbeit (GIZ) in the northern part of country between 1984 and 1995, today its spread is limited to a handful of farmers – as testified to by the experts of the multi-stakeholder meeting under this recent study. This seems to be in contrast to the adoption of conventional narrow-based tied ridges (e.g. Mutekwa et al., 2005) and scientific work pointing to their positive impact on *in situ* soil moisture retention. For example, Motsi et al. (2004) conducted an experiment with four water conservation treatments (tied ridges, infiltration pits, *fanya juu* terraces and conventional ploughing on the flat as the control) in low-rainfall areas (<500 mm per year). Their results showed that tied ridges were ranked best by both the stakeholders' evaluation and the soil moisture measurements (i.e. theta (h)-probe readings) in retaining moisture compared to the other treatments. The farmers who practised tied ridges realized yields of about 3 tonnes/ha compared to conventional tillage treatments whose yields were about 1.5 tonnes/ha. Being a narrow-based (rather than broad-based) ridge and furrow system, the results are related to an *in situ* rather than inter-field water management effect: the ridges were constructed with a row spacing of 0.75 m (rather than 1.50 m in the case of tied furrows), followed by:

> an operation to tie the ridges before planting, using hand hoes, with small mounds along each furrow so as to impede the runoff of the rainwater. The mounds were at intervals between 2 and 3 m and care being taken to leave them at a height that was less than that of the main ridge to be used for planting.
>
> Motsi et al. (2004)

Between 6 and 15 January 2012 a multiple-stakeholder meeting was held in Harare and a revisit of the water harvesting sites was conducted in Zimbabwe. The stakeholders and experts present at the meeting were drawn from NGOs, the Department of Agricultural, Technical and Extension Services (AGRITEX), University of Zimbabwe and the Ministry of Agriculture, Mechanization and Irrigation, Department Soil & Water Conservation. Reasons mentioned by experts for abandoning the broad-based tied furrow system include:

- The need for equipment and tools, lack of (donor) funding, lack of upscaling activities and uncertainty about how the technology would work in different environments;
- Donor funding, if any, that is targeting conservation agriculture[3] and not water catchment systems;
- The tied furrow system was particularly promoted through projects, hence, practised at certain locations (and not spreading spontaneously) and abandoned when projects move out;
- With the start of the economic crisis from 2000 onwards, there were changes in ministries, lack of funds and lack of motivation among extension workers to upscale the technology;
- Since 2000, there has been a lot of uncoordinated implementation of the technology among the NGOs. Tied ridging and furrowing is considered as a technology not fitting conservation agriculture, and, hence, it is discouraged and kept outside funding programmes, if any;
- NGOs promote *in situ* technologies and not water harvesting like the tied furrow system because of limited access to equipment required to construct

the ridges and furrows. Farmers are therefore forced to make ties by hand making it labour intensive. This has contributed to the technology being abandoned completely;

- The structures made in the dry season were damaged by grazing cattle or (partly) washed away in the wet season.

Dead-level contour

Instead of diverting water away from the field, as is done by standard graded contours, the purpose of the dead-level contour is to collect runoff water from outside and within the field and store it in the channel and the infiltration pits. The water will then slowly percolate into the soil on either side of the contour, providing vital moisture to the adjacent crops (where cropping density can be higher, and crops more varied) during dry spells. The reasons for this modification have been manifold; the principle aim being to improve crop productivity – especially cereals – through better moisture retention technologies, and thereby

- produce enough food that last for the whole year even after experiencing long dry spells;
- increase household assets; and
- improve well-being of household members.

Design and technology innovation

The innovative design of the dead-level contours includes the construction of zero-gradient channels upslope of the earth ridge (minimum size: 50–125 cm deep, 150 cm wide, 100–200 m long) which often contain infiltration pits or storage tanks (usual size: 125 cm deep, 150 cm wide and 200 cm long; storage capacity: 2.50 to 3.75 m^3 on average) as shown in Figure 10.4.

The horizontal distance (spacing) between the channels averages 12 m (for slopes ≤ 2 per cent) and that for the pits within the channels between 5 and 25 m, depending on soil type (with usually a wider spacing for sandy soils). The depths of channels and pits in medium-textured soils are greater than those in sandy soils, to allow for higher water storage and compensate for the soils' lower infiltration rates. In light-textured sandy soils the bases of the pits are compacted or rammed, using soil from elsewhere (e.g. soil from termite mounds) to prevent deep percolation. The infiltration pits in the drier areas (mean annual rainfall ≤ 500 mm) have a temporary cover to reduce the loss of water through evaporation. A pit cover is constructed from wooden poles placed across the pit which are then stabilized by placing hessian bags and earth on top. The soil derived from excavating the channels and pits is used to make bunds along the downslope sides of the channels. The bunds, usually covered after a while with natural (non-planted) grasses, avoid overtopping of excess water.

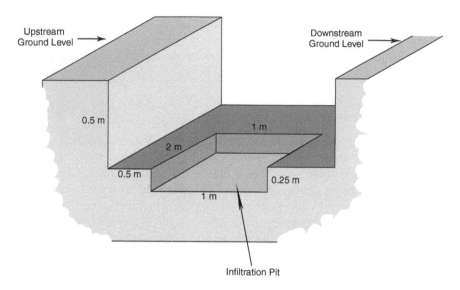

Figure 10.4 Cross-section of infiltration pit within dead-level contour

Establishment and maintenance

The implementation of dead-level contours on farms encompasses a series of activities. After gathering the materials for construction (e.g. A-frame,[4] pegs, rammer, poles, pick, shovel, hacksaw, axe), the farmer marks the location of the contour channels in the field using an A-frame. Then, with the assistance of hired labourers and/or family members, the channels are excavated with shovels and picks. If farmers observe that the DLCs quickly lose the harvested rain and runoff water due to evaporation and deep percolation, they then construct infiltration pits along the channels (total labour requirement: 80 person days amounting to US$400 for digging a 200-m-long contour channel with infiltration pits in medium to heavy-textured soils; half the time and costs for light-textured soils; on average four to five contour channels of 150–200 m are established on a field of 50–100 m by 100m). In light-textured soils, the pit base is rammed with soil gathered from elsewhere (ca. 3 m^3 of termite soil; i.e. three half-full carts, is needed for ramming all pits along five 200-m-long contour channels; total costs are US$15 for the carts and additional US$25 for five person days of soil digging and application). In the drier areas a temporary cover is constructed and placed over the pit. During the dry season when all water has been used from the storage facilities (usually soon after harvesting), the temporary cover is removed from the pit. The soil deposits and plant debris are removed from the pit and channel sections and spread again in the field where the material came from. The soil and plant trash form a good nutrient source for the crops. The sides of the storage facilities (channels or pits)

Photo 10.1 A dead-level contour with a covered infiltration pit. Matabeleland South, Gwanda-Manama (D. Snelder)

are repaired and re-plastered where needed. Repairs are particularly essential at spots where water seepage has been observed, allowing too much water to pass through. Crop residues including maize stalks, weeds and grasses, rotten fruits and vegetables (e.g. melons and pumpkins) and other waste material are stacked into the open storage facilities, serving as compost to be applied to the fields for maintenance of soil fertility at the start of the rainy season. After the first rains, when all compost has been removed from the storage facilities, the pits are covered again and the plastering work completed to ensure optimal harvesting of rainwater and runoff. For the maintenance activities (labour: maximum of one person day per 100 m contour channel), the working in groups – preferably the same groups that dug the channels and pits before – is preferred, allowing the sharing of knowledge and experience while working. The inputs are thus mainly in the form of labour, with labour demands being highest during establishment (excavation and digging activities) and pay off after a longer time. The total labour cost for establishment of a 150 m DLC using 2012 prices is estimated at around US$54 and US$37.50 for yearly maintenance after first year of establishment.

Cropping system and human environment

The annual crops grown in NR IV and NR V (rainfall less than 600 mm/year) include maize, sorghum and millets and the intercrops include groundnuts,

Photo 10.2 Dead-level contour (left of photo) in a recently planted maize field, Zvishavane (D. Gumbo)

watermelon and pumpkin. The conversion of standard graded contours into dead-level contours has resulted in higher yields; that is, average maize yields are 0.6–0.8 tonnes/ha for fields with standard graded contours and 1.5–2.5 tonnes/ha for fields with DLCs; for sorghum, the average yields are 0.8 tonnes/ha with standard graded contours but considerably more (up to 4.0 tonnes/ha) with DLCs. Droughts, however, regularly affect crop yields at a frequency that has increased in recent decades. During the recent site revisit, farmers' observations were that in the past, one good harvest year was followed by three less good years, whereas at present it is one good harvest year followed by five less good years. The DLCs are mainly implemented by small-scale farmers (of 'average wealth') to produce the short-seasonal food crops for their individual households and selling at local markets. The grass growing spontaneously on the contour ridges along the channels is used as fodder for livestock. Animals are allowed to graze in the fields after harvesting. Most farmers keep livestock (up to 6 cattle, 10 goats and/or sheep and 2–4 donkeys for the average farmer). The land being used for DLCs is either untitled individual land or communal (village) land. The size of cropland per household varies from 0.5 to 1 ha (average farmers) up to 2–5 ha (rich farmers). Ten to fifty per cent of the farmers using DLCs earn a side income from gold panning and remittances (money sent by relatives living abroad).

Upscaling

Up to 1988, the innovations and adjustments of the standard graded contours (leading to dead-level contours) were farmer-driven. In 1988, about ten farmer innovators in Zvishavane and Chivi districts were part of the Indigenous Soil and Water Conservation in Africa Project to share their knowledge and discuss DLC innovations (Hagmann and Murwira, 1996). After that, the technology of DLCs was spread, and the number of adopters increased to about 5000 in 10 pilot districts.

The spread was mainly through exposure and farmer-to-farmer visits organized by NGOs (e.g. Africa 2000, CARE, Christian Care, Lutheran Development Service Practical Action, World Vision International, Zvishavane Water Project) and supported by AGRITEX and other international programmes (e.g. FAO programmes, World Food Programme's Food for Assets). However, from 2000 onwards, the partnerships have been affected by the economic crisis and political turmoil. Extension workers lack funds to provide support services to farmers, making them look for better-paid jobs abroad. Moreover, AGRITEX has been split into different departments, with the soil and water conservation unit being shifted to the department of mechanization. The latter department has only staff at provincial and not at district level, hampering access by farmers who are faced with high transport costs. In addition, farmers have been resettled on land without secure tenure arrangements and, hence, lack the necessary incentives to invest. Currently, there are only few NGO and government initiatives; most farmers are mainly experimenting on their own. However, for the projects previously supported by NGOs, about 40 per cent of the farmers continue to use and maintain the DLC technology. The main reason for the continued use of the technology is the benefits of better crop performance during the dry spells.

In the past promotion and extension of soil and water technologies (namely standard graded contours) was through legislation and coercion. At present, various approaches towards up-scaling, innovation and extension of the improved soil and water technologies are being used. Some farmers, particularly those living in the drier regions of Zimbabwe, conduct on-farm experimentation by themselves, modifying and innovating the DLCs by trial and error. NGOs and extension workers promote DLCs through on-farm demonstrations and exposure visits. International NGOs promote DLCs using subsidies and incentives. A good example is the food-for-asset initiative, the asset being the DLC while food is used as payment for creating the DLCs in one's field. Partnerships in research are also set up between farmers, researchers and extension workers. Adaptation of technologies is greatly improved through on-farm research where one works with innovators rather than working with a team of scientists only.

Constraints

Most land users do not have adequate tools and equipment to construct the dead-level contour channels. The pick and shovel cost between US$10 and US$12 each, which is an expenditure that most poor smallholder farmers cannot afford. Those who manage to buy tools often complain of tools getting worn out even before finishing constructing the channels. Some land users have tried to access cheaper tools from local blacksmiths. In addition, extension officers have not been able to support farmers with technical advice on the DLC technology as they are either immobile (lack of transport) or, due to the harsh economic conditions, lack of motivation to go and assist the farmers. Likewise, whereas the farmer-to-farmer approach, where farmers learn from each other, has been introduced during training, there is limited tracking (monitoring) to establish the effectiveness of the approach.

There are no technical guidelines or 'how to' manuals that farmers can use as reference material when constructing the DLCs. The only guidelines are in the form of technical briefs which are solely in English and not in a language most farmers can easily understand (Practical Action, 2012). The DLC technology demands considerable (non-skilled) labour in digging the channels and pits. Forming groups of 10 to 15 people has helped some land users to share labour. Hired labour has also been used to complement family labour.

Some land users have used ox-drawn ploughs to dig the first 15–20 cm of the contour channels and finished the required depth of 50 cm by using family or hired labour. Since 2000, a large number of men have left their villages to work in towns often outside the country, leaving the task to women. In addition, HIV/AIDS has caused a significant loss in the labour supply, particularly during the period 2000–5.

The right time to construct the DLC (soon after harvesting, when the soil is still a little moist) often coincides with the communal grazing of fields resulting in damaged structures and, hence, high maintenance costs. When construction is done later in the dry season, the soil is often too dry and hard, making it difficult to dig. Fencing is very expensive for poor-resource farmers, with fences being often vandalized during the communal grazing period. Farmers have therefore often invested in the DLC technology for the small fields around their homesteads which are usually fenced and easier to supervise.

Before 1988, the land tenure system was usufruct. After the fast track land reform that reached its peak in 2000, land users in new resettlement areas are not willing to invest in DLCs – even those who have the knowledge of the technology – because they lack land tenure security (see Zimbabwe: background). Moreover, the land in communal areas, where titles are secured, is often left fallow or contains poorly maintained DLCs. The same people who are mentioned to have been resettled are still holding on to their previous plots in communal areas, suggesting they have not totally moved to the new resettlement areas. In addition, they lack sufficient labour to work in both areas resulting in poor maintenance of the DLC and even leaving whole fields fallow.

Conclusions and recommendations

Whereas water harvesting for crop production was traditionally non-existent in the drier areas of Zimbabwe, in recent decades various developments have been made in this direction. The tied furrow system and the dead-level contours (DLCs) are examples, yet the former has virtually been abandoned after a period of experimentation on research stations and farms in the 1980s and early 1990s. Farmer-based innovations and modifications to the DLC have resulted in significantly lower water losses and higher soil moisture storage which has improved crop condition and biomass during intra-seasonal dry spells. Adaptation of the technology is greatly improved through on-farm research where one works with innovators rather than working with a team of scientists only. The uptake and upscaling of DLCs by smallholder farmers and support organizations can be improved by increasing the number of knowledge sharing platforms at community and district levels. More research into the feasibility of DLCs for different cropping systems and soils is ongoing. Research is required to give farmers and other stakeholders clear guidelines on water-harvesting technologies suitable for different climatic and soil conditions. The research should provide biophysical information on where the technology systems (tied furrows, DLCs) are applicable. The research needs also to focus on the whole catchment area rather than being limited to the field or farm in which the technology is practiced.

Notes

1 These are technically known as 'channel terraces with waterways'.
2 The word 'ridge' in Zimbabwe is referred to as a 'bund' elsewhere in Africa.
3 Conservation agriculture (CA) involves cultural practices that conserve soils, extend the period of water availability and improve soil fertility. CA is based on three pillars namely: minimum soil disturbance, crop rotation and maintenance of permanent soil cover.
4 An A-frame is a simple device made by tying tightly three wooden sticks (two of the same length – about 2 m long and the third half the length of the two longer sticks) to form a capital 'A' and is used to find points across a slope that are at exactly the same level. The level points are located by freely hanging with a string, a small rock with the other end tied to the top of the 'A' frame.

References

Aylen, D. (1941) 'Who built the first Contour Ridges', *The Rhodesia Agricultural Journal*, vol. 36, pp. 452–484.
Alvord, E. D. (1958) *Development of Native Agriculture and Land Tenure in Southern Rhodesia*, Unpublished manuscript, Rhodes House, Oxford.

Anderson, I. P., Brinn, P. J., Moyo, M. and Nyamwanza, B. (1993) 'Physical resource inventory of the Communal lands of Zimbabwe – An overview', *NRI Bulletin 60*, Natural Resource Institute, Chatham, UK.

Bird, K., and Busse, S. (2007) 'Re-thinking aid policy in response to Zimbabwe's protracted crisis: A Discussion Paper', ODI Discussion Paper.

Bird, K. and Shepherd, A. (2003) 'Livelihoods and chronic poverty in semi-arid Zimbabwe', *World Development*, vol. 31, no. 3, pp. 591–610.

Birner, R. and Resnick, D. (2010) 'The political economy of policies for smallholder agriculture', *World Development*, vol. 38, no. 10, pp. 1442–1452.

Bratton, M. (1987) 'Drought, food and social organisations of small farmers in Zimbabwe', in: M. H. Glantz (ed.), *Drought and Hunger in Africa: Denying famine a future*, Cambridge University Press, Cambridge.

Chimhowu, A. and Hulme, D. (2006) 'Livelihood dynamics in planned and spontaneous resettlement in Hurungwe District, Zimbabwe: converging and vulnerable', *World Development*, vol. 34, no. 4, pp. 728–750.

Chipika, J. T. and Kowero, G. (2000) 'Deforestation of woodlands in communal areas of Zimbabwe: is it due to agricultural policies?', *Agriculture, Ecosystems & Environment*, vol. 79, no. 2, pp. 175–185.

Chuma, E., Chiduza, C. and Utete, D. (1997) *Training Manual: Soil fertility management for smallholder farmers in Zimbabwe*. Institute of Environmental Studies, University of Zimbabwe, Harare, Zimbabwe.

Critchley, W., Reij, C. and Seznec, A. (1992) 'Water harvesting for plant production. Volume II: Case studies and conclusions for Sub-Saharan Africa'. World Bank Technical Paper Number 157. Africa Technical Department, World Bank, Washington D.C.

Dreyer, S. (1997) 'It's just like growing shallots ...! Experiences with vetiver grass in a soil and water conservation programme for communal farmers in Zaka District, Zimbabwe', available at: www.vetiver.org/ZIM_zaka.htm, accessed 29 May 2012.

Elliot, J. (1987) 'Soil conservation in Zimbabwe: Past policies and future direction', *Geographical Journal of Zimbabwe*, vol. 18, pp. 53–67.

FAO (2006) Country Pasture/Forage Resource Profiles, Zimbabwe, Gambiza, J. and Nyama, C (eds), available at: http://www.fao.org/ag/AGP/AGPC/doc/Counprof/zimbabwe/zimbab.htm, accessed 29 May 2012.

FAOSTAT (2011) Country profiles – Zimbabwe, available at: http://faostat.fao.org/site/666/ default.aspx, accessed 30 April 2012.

Hagmann, J. and Murwira, K. (1996) 'Indigenous soil and water conservation in southern Zimbabwe: a study of techniques, historical changes and recent developments under participatory research and extension', in: C. Reij, I. Scoones and C. Toulmin (eds), *Sustaining the Soil. Indigenous Soil and Water Conservation in Africa*, Earthscan, London.

Kahinda Mwenge, J-M., Rockström, J., Taigbenu, A. E. and Dimes, J. (2007) 'Rainwater harvesting to enhance water productivity of rainfed agriculture in the semi-arid Zimbabwe', *Physics and Chemistry of the Earth*, vol. 32, no. 15–18, pp. 1068–1073.

Kronen, M. (1994) 'Water harvesting and conservation techniques for small holder crop production systems', *Journal of Soil and Tillage Research*, vol. 32, pp. 71–76.

Kwashirai, V. C. (2006) 'Dilemmas in conservationism in colonial Zimbabwe 1890–1990', *Conservation and Society*, vol. 4, no. 4, pp. 541–561.

Mapedza, E. (2007) *Keeping campfire going: political uncertainty and natural resource management in Zimbabwe*. Gatekeeper Series no. 133, London: IIED.

Motsi, K. E., Chuma, E. and Mukamuri, B. B. (2004) 'Rainwater harvesting for sustainable agriculture in communal lands of Zimbabwe', *Physics and Chemistry of the Earth*, vol. 29, pp. 1069–1073.

Mugabe, F. T. (2004) 'Evaluation of the benefits of infiltration pits on soil moisture in semi-arid Zimbabwe', *Journal of Agronomy*, vol. 3, no. 3, pp. 188–190.

Munamati, M. and Nyagumbo, I. (2010) 'In situ rainwater harvesting using dead-level contours in semi-arid southern Zimbabwe: Insights on the role of socio-economic factors on performance and effectiveness in Gwanda District', *Physics and Chemistry of the Earth*, vol. 35, pp. 699–705.

Mupangwa, W., Love, D. and Tomlow, S. (2006) 'Soil-water conservation and rainwater harvesting strategies in the semi-arid Mzingwane catchment, Limpopo Basin, Zimbabwe', *Physics and Chemistry of the Earth*, vol. 31, pp. 893–900.

Mupangwa, W., Tomlow, S. and Walker, S. (2011) 'Dead-level contours and infiltration pits for risk mitigation in smallholder cropping systems of southern Zimbabwe', *Physics and Chemistry of the Earth*, Parts A/B/C, available online 8 July 2011, DOI: 10.1016/j.pce.2011.06.011.

Mutekwa, V. and Kusangaya, S. (2006) 'Contribution of rainwater harvesting technologies to rural livelihoods in Zimbabwe: The case of Ngundu ward in Chivi District', *Water SA*, vol. 32, no. 3, pp. 437–444.

Mutekwa, V., Kusangaya, S. and Chikanda, A. (2005) 'The adoption of rainwater harvesting techniques in Zimbabwe, the case of Chivi ward in Masvingo', available at: bscw.ihe.nl/pub/bscw.cgi/.../Kusangayapercent202.pdf, accessed 29 May 2012.

Nyagumbo, I., Munamati, M., Chikwari, D. E. and Gumbo, D. (2009) 'In-situ water harvesting technologies in semi-arid southern Zimbabwe: Part I. The role of biophysical factors on performance in Gwanda district', Paper presented at the 10th WaterNet/WARFSA/GWP-SA Annual Symposium, 28–30 October in Entebbe, Uganda.

Nyagumbo, I. and Rurinda, J. (2011) 'An appraisal of policies and institutional frameworks impacting on smallholder agricultural water management in Zimbabwe', *Physics and Chemistry of the Earth*, vol. 47–48, pp. 21–32, available online 27 July 2011, DOI:10.1016/j.pce.2011.07.001.

Nyamapfene, K. (1991) *Soils of Zimbabwe*, Nehanda Publishers, Harare, Zimbabwe.

Nzuma, J. K. and Murwira, H. K. (2000) 'Improving the management of manure in Zimbabwe', *Managing Africa's Soils*, no. 15, p. 20.

Phillips, J. G., Cane, M. A. and Rosenzweig, C. (1998) 'ENSO, seasonal rainfall patterns and simulated maize yield variability in Zimbabwe', *Agricultural and Forest Meteorology*, vol. 90, no. 1, pp. 39–50.

Practical Action (2012) 'Dead-level contours', available at: http://www.appropedia.org/Dead_level_contours, accessed 29 May 2012.

Scoones, I., Chibudu C., Chikura, S., Jerenyama, P., Machaka, D., Machanja, W., Mavedzenga, B., Mombeshora, B., Mudhara, M., Mudziwo, C., Murimbarimba, F. and Zireza, B. (1996) *Hazards and Opportunities. Farming Livelihoods in Dryland Africa: Lessons from Zimbabwe*, Zed Books, London.

Sepaskhah, A. R. and Fooladmand, H. R. (2004) 'A computer model for design of micro catchment water harvesting systems for rain-fed vineyard', *Agricultural Water Management*, vol. 64, pp. 213–232.

Torrance, J. D. (1981) *Climate Handbook of Zimbabwe*, Zimbabwe Department of Meteorological Services, Harare.

UNDP (2010) *Zimbabwe: Coping with drought and climate change*, available at: http://www.undp.org/gef/adaptation/projects/CwD/ or http://www.adaptation learning. net/projects/zimbabwe-coping-drought-and-climate-change, accessed 29 May 2012.

Unganai, L. S. (1996) 'Historic and future climatic change in Zimbabwe', *Climate Research*, vol. 6, pp. 137–145.

Vincent, V. and Thomas, R. G. (1960) *An agricultural survey of Southern Rhodesia: Part I: agro-ecological survey*, Government Printer, Salisbury.

Walker, S., Tsubo, M. and Hensley, M. (2005) 'Quantifying risk for water harvesting under semi-arid conditions. Part II Crop yield simulation', *Agricultural Water Management*, vol. 76, no. 2, pp. 94–107.

World Bank (2009) *2009 World Development Indicators*, The World Bank, Washington D.C.

Wilson, K. (1988) 'Indigenous conservation in Zimbabwe: soil erosion, land use planning and rural life', Paper presented at the African Studies Association Conference, September 1998, Cambridge.

Chapter 11

Investing in water for agriculture in the drylands of Sub-Saharan Africa

Considerations for a conducive policy environment

Denyse Snelder, Martin Bwalya, Hilmy Sally and Maimbo Malesu

Introduction

Both public and private investment in water for agriculture as a means of poverty reduction require an enabling policy environment with supporting, socially embedded water institutions, particularly where it concerns smallholder farmers in the drylands of Sub-Saharan Africa. Longer-term policies with integrated management perspectives will facilitate technology uptake particularly where combined with multiple-use systems. The same is true of the need for policies conducive to the private sector for investing in innovative water technologies in a cost-effective way (Molden, 2007).

Whereas fear of famine among the rural poor was the major driving force behind water-related agricultural investments in developing countries in the past (de Fraiture et al., 2010), current investments are increasingly driven by the need to respond to changes in diet and energy sources creating associated demands for meat, high-value food and biofuel crops by a rapidly increasing world population (de Fraiture and Wichelns, 2010). Although smallholder agriculture remains a key option in reducing poverty and hunger (Hazell et al., 2010), the future of small farms in Africa largely depends on government policy and investment decisions (Jayne et al., 2010) and the much needed institutional innovations to overcome market failures (Birner and Resnick, 2010; Hazell et al., 2010). Government policies have often been unfavourable or contradictory with regard to investment in smaller scale water technologies for agricultural production purposes. When (donor) funds become available, governments tend to favour highly-visible large-scale infrastructure investments for water management despite the fact that small-scale technologies, like water harvesting, offer a faster and more cost-effective way to economically empower rural farming communities thereby contributing to achieving the Millennium Development Goals. Water management institutions may further need more flexibility to change policies as the understanding of drivers and associated effects improves. On the other hand, multi-stakeholder coordination and negotiation mechanisms are required to deal with tradeoffs and innovative ways of implementing decisions (Inocencio et al., 2007). The poorest stakeholder groups generally have little voice and political power to realize access to water and capital and negotiate allocations for water (WRI, 2005), suggesting a need for integrating capacity building into local policies.

The objective of this chapter is to address the question whether Sub-Saharan Africa has the appropriate policy environment to stimulate and facilitate increased investments in water for agriculture. After an introduction to the trends and challenges in food demand and consumption, poverty and water use in agriculture in the following section, the chapter will discuss different aspects related to the question above. It will highlight policy issues and policy design considerations related to water resource management and investment with the ultimate goal of rising agriculture productivity and improving cropping stability in Sub-Saharan Africa. Special attention will be paid to water harvesting systems in the rainfed drylands; that is, the focus of this book.

Trends and challenges

Food demand, consumption and production

Growth in food demand is highest in developing countries where populations are growing rapidly and per capita income is low but increasing, thus allowing a greater part to be used for improving daily diets. Although growing at a slower pace compared to previous decades, the population in Sub-Saharan Africa is still increasing at a rate well in excess of 2 per cent compared to 0.2 per cent in developed countries. The income growth is also relatively strong (in 2010: 5 per cent annual growth, compared to 3 per cent for EU – in adjusted net national income; i.e., GNI minus consumption of fixed capital and natural resources depletion) but unequally distributed across the population of the region and hence moderating growth in per capita food consumption like in previous decades (OECD-FAO, 2011). This trend is still in contrast with the stagnant or downward trend in food consumption growth in high-income countries where markets are saturated (OECD-FAO, 2011).

Total cereal production (in million tonnes) in Sub-Saharan Africa steadily increased over past decades; that is, 37.8 in 1970, 42.8 in 1980, 58.6 in 1990, 73.2 in 2000, 83.2 in 2003 and 121.3 in 2009, whereas the projected production for 2030 is 168 million tonnes (FAO, 2003, 2005, 2010a). The region's cereal yield was 1295 kg/ha in 2009, however, rates varied considerably among countries, with lowest yields recorded for Niger (380 kg/ha) and Zimbabwe (450 kg/ha) and highest for South Africa (4414 kg/ha) (World Bank Database, 2012). The total cereal production has not been sufficient to meet domestic demands as is evident from increasing cereal imports over past decades: a total of 28.9 million tonnes of cereals was imported into the Sub-Saharan Africa region in 2009, compared to 9.6 million in 1990 and 8.5 million in 1980 (FAO, 2012). Imports of Kenya, Ethiopia and South Africa amounted to 2.7, 2.2 and 2.1 million tonnes respectively in 2009 (FAO, 2012).

Poverty and undernourishment

Inadequate food production, high food prices and the global economic crisis during the past decade have threatened the world's ability to achieve internationally agreed goals on hunger reduction: the first Millennium Development Goal (MDG) and the 1996 World Food Summit goal. Whereas most (537.2 million) of the world's 739.2 million hungry people live in the world's most populous region of Asia, Sub-Saharan Africa is home to 160.3 million of the world's undernourished population (22 per cent of the world's total, 23 per cent of the Sub-Saharan Africa population; 2006–8 figures; FAO, 2012); that is, an annual increase of 0.1 per cent when compared to 159.1 million in 1995–97 (31 per cent of the Sub-Saharan Africa population; FAO, 2012). Yet, large variations occur among countries with the highest numbers recorded for Ethiopia (32.6 million), Tanzania (13.9 million) and Kenya (12.4 million; 2006–8 figures, FAO, 2012). With 72 per cent of the population living below US$2 per day (2000/2009 data; PRB, 2011), compared to 54 per cent for Asia (64 excluding China; PRB, 2011), it is unlikely that even economic growth, while essential, will be sufficient in itself to eliminate hunger within an acceptable period of time (FAO, 2010b, 2012).

Use of water in agriculture

International institutions including the World Bank report that over 40 per cent of the extra food needed to meet increasing food demands in the coming years will have to come from intensified rainfed farming (World Bank, 2006). The FAO estimated that cereal production in Sub-Saharan Africa would need to increase at about 2.6 per cent per year over the next two decades (compared to an annual increase of about 1.7 per cent or less for other regions of developing countries, with average of 1.3 per cent for all developing countries; FAO, 2005). This stresses the need to find ways to accelerate agricultural growth in Sub-Saharan Africa. Policies – both international and national – will have to focus on investments and incentive packages needed to prompt such a rapid rate of growth. The role of water in enhancing agricultural production is one of the key aspects to be addressed in such policies to ensure this fast growth. Molden et al. (2010) identified priority areas where substantive increases in water productivity can be expected as those areas where poverty is high and water productivity is low, where water is scarce and competition for water is high, and where a little extra water use through water resource development can make a big difference. The water resource endowment and water productivity in various countries in Sub-Saharan Africa suggests that the region is among one of those priority areas.

Worldwide, rainfed agriculture covers 80 per cent of the agricultural cropland (total agricultural land area in 2009: 48.8 million km^2) and produces 60 per cent of the crops, whereas irrigated agriculture covers 20 per cent of all cropland (in FAO terms: arable land and permanent crops) and is estimated to produce 40 per cent of all crops (FAO, 2010, 2012). Of the total 10.5 million km^2 agricultural land in Sub-Saharan Africa, only 3 per cent is devoted to irrigation (2008 data in FAO 2012), the rest being rainfed. Hence, water withdrawals for agriculture are very limited – just under two per cent of the total renewable water resource – and water storage is well below levels in other regions.

Table 11.1 shows the per capita green, blue and grey water footprint (WF) of production and consumption for Sub-Saharan Africa and various countries in the region. The water footprints refer to the consumptive use of rainwater (green WF) and ground and surface water (blue WF) and the volume of freshwater that is required to assimilate the load of pollutants based on existing ambient water quality standards (grey WF; Hoekstra and Mekonnen, 2012; see also Chapter 2). Agriculture accounts for 95 per cent of the total blue water footprint in Sub-Saharan Africa, amounting to 31.580 Mm^3/year, the remainder shared between industrial production (1 per cent) and domestic water supply (4 per cent). Whereas withdrawal of water for industrial and domestic purposes is expected to be more or less equally distributed throughout the year, withdrawal of water in the agricultural sector is variable and related to irrigation practices during the crop growing season. The use of water in agriculture also leads to the release of significant volumes of polluted water: agriculture accounts for 46 per cent of the total grey water footprint (related to pollutants released in the production of crops – like maize and wheat and also poultry, in the case of South Africa), followed by industrial production with 40 per cent and domestic supply with 14 per cent. Among the countries listed in Table 11.1, Sudan shows the highest per capita blue water footprint which is particularly related to the consumption of agricultural products such as milk, sorghum, sugar and wheat. Most of the water footprint is in the form of irrigated water supply, which does not come as a surprise, given the country's arid to semi-arid climate (see rainfall data in Table 11.1). The green water footprint is by far the highest footprint in all countries, with Niger showing a per capita footprint that is two to three times the amounts of the other countries. The high level is completely accounted for by the high consumptive use of rainwater in the agricultural sector; that is, for the production of millet, pulses, meat and milk in the case of Niger. Under current (and future) conditions of climate change, the water use in the agricultural sector is expected to rise even further to meet the increasing food demand and need to raise crop production. Moreover, with climate change projections pointing at higher variability in annual rainfall quantities, intensities, and incidences of drought and flooding, crop failure will increase and lead to replanting activities to make up for lost yields resulting in yet more water use.

Investing in water for agriculture

Rainfed agriculture currently plays the dominant role in producing the world's food supply: about 62 per cent of the gross value of the world's food is produced under rainfed conditions on 80 per cent of the world's cropland (FAO 2012). The World Bank even expects that over 40 per cent of the extra food needed to meet

Table 11.1 The water footprint of national consumption per capita during 1996–2005 period for selected countries in Sub-Saharan Africa (m³/year/cap)

Country	Rainfall (mm/year)[a]	Population[b] (million)	Water footprint of consumption of agricultural products[b]			Water footprint of consumption of industrial products[b]		Water footprint of domestic water consumption[b]		Total water footprint of national consumption[b]			Ratio external/total water footprint (%)[b]
			Green	Blue	Grey	Blue	Grey	Blue	Grey	Green	Blue	Grey	
Burkina Faso	748	11.9	1637.8	30.2	24.2	0.2	2.2	0.9	7.8	1637.8	31.2	34.2	3.4
Ethiopia	848	66.5	1128.1	24.5	8.2	0.1	1.3	0.5	4.5	1128.1	25.1	14.0	2.3
Kenya	630	31.9	1031.0	32.6	17.2	0.4	5.6	1.5	13.1	1031.0	34.4	35.9	17.4
Niger	151	11.2	3411.1	86.3	12.3	0.1	1.0	0.8	7.2	3411.1	87.1	20.5	2.5
Rwanda	1212	7.7	799.6	7.8	6.6	0.1	2.2	0.5	4.2	799.6	8.4	13.0	4.5
South Africa	495	45.1	1027.3	87.4	62.4	1.6	15.6	8.6	52.4	1027.3	97.6	130.4	22.0
Sudan	416	35.2	1484.5	200.9	14.6	0.4	7.2	2.8	25.3	1484.5	204.1	47.1	3.9
Tanzania	1042	34.7	966.3	32.9	9.3	0.2	2.3	1.5	13.7	966.3	34.6	25.3	6.8
Uganda	1180	24.9	1055.1	10.3	4.6	0.2	3.6	0.5	4.7	1055.1	11.0	12.9	4.7
Zambia	1020	10.5	830.2	35.7	20.9	0.4	6.6	2.7	24.7	830.2	38.9	52.2	N/A
Zimbabwe	657	12.3	1039.8	64.3	40.1	1.0	17.4	4.8	42.9	1039.8	70.1	100.4	7.9
SSA total (Mm³/year)		675	774,594	31,580	11,482	257	3,413	1,252	10,088	774,594	33,089	24,983	8.4

a: Source: FAO-AQUASTAT (http://www.fao.org/nr/water/aquastat/main/index.stm); Tanzania data: SUA 2007, (www.suanet.ac.tz/ccaa/downloads/crTanzania.doc)
b: Source: Mekonnen and Hoekstra (2011); population numbers are 2005 data.

increasing food demands in the coming years will have to come from intensified rainfed farming (World Bank, 2006). The proportion of future food production that could or should come from rainfed or irrigated agriculture is, however, subject to debate. In Sub-Saharan Africa a total 7.1 million ha is equipped for irrigation (2008 data in FAO 2012), which is just over two per cent of the world's total area equipped for this purpose (304.4 million ha); that is, a rather limited area when compared to the 218.9 million ha in Asia (72 per cent of world total; 2008 data in FAO 2012). Moreover, actual yields are low but this offers a large, untapped potential to increase productivity. The potential for irrigation in Sub-Saharan Africa is estimated at 39.6 million ha; that is, less than 4 per cent (based on FAO 2012 data) of the region's total agricultural land area and accounting for even less than 1 per cent of the world's total agricultural area (based on FAO 2012 data). Hence, the bridging of yield gaps (i.e. gaps between actual and potential yields) is expected to have a relatively small impact on the total food supply, as the total irrigated area is, and will be, relatively small (de Fraiture and Wichelns, 2010). Moreover, expanding irrigated areas to reach the irrigation potential will require investments in associated inputs, infrastructure, and market access to maximize agricultural productivity. The poverty reducing impacts of irrigation water are reported to be greatest when such combined investments are met (Hanjra et al., 2009; Namara et al., 2010), which is in fact also true for poverty reduction in areas depending on rainfed agriculture. Livelihoods might be improved more effectively by investing in activities that upgrade production in rainfed agriculture, taking into account that millions of poor farmers in Sub-Saharan Africa depend on rainfed agriculture. The latter is far more extensive in the area it covers compared to irrigated agriculture (see also Molden, 2007; de Fraiture et al., 2010).

To date, there has been less agricultural water development in Sub-Saharan Africa than in any other region, with investment costs in the rainfed areas tending to be lower than in irrigated agriculture (Molden, 2007). A report (World Bank, 2008) based on a collaborative programme on investment in agricultural water for poverty reduction and economic growth in Sub-Saharan Africa, points to the neglect of water-in-agriculture development in investment programmes by stating:

> Early poverty reduction strategy programmmes did not always explicitly recognize the critical role of the agriculture sector in poverty reduction and growth, although more recent examples have done so. They have, however, generally still not assigned much prominence to agricultural water development. Consequently, the subsector has tended to be neglected in investment programmes for the agriculture and water sectors.

Potential reasons for this neglect are the negative perceptions of agricultural water management. This is due to high cost and uneconomic returns, particularly large-scale investments in irrigation projects in the 1970s and 1980s, viewed as projects with greater environmental and social risk. In addition, in most countries, there are various ministries involved in water and agriculture leading to divided responsibility (IFAD, 2002; World Bank, 2008).

Levels and trends of donor financing are often taken as a proxy for investment levels. Water for agriculture is currently financed from a mixture of internal and external assistance sources, including foreign bilateral and multi-lateral donors, national and local governments, public agencies, private institutions and farmers at all scales and varying types. It has been estimated that the annual costs of managing water resources to meet the MDG hunger goals is in the order of US$45–50 billion over the next ten years rising to US$65–70 billion thereafter (WG, 2006). The prospects for a major expansion of government subsidies are however not promising, given fiscal barriers and multiple competing claims on public funds.

The per capita net Official Development Assistance (ODA) received in 2009 in Sub-Saharan Africa region is US$47.1, varying between US$20 and US$100 among the selected countries in Table 11.2. Foreign investment in Sub-Saharan Africa increased significantly during the 1995–2010 period: the inflow of foreign direct investment in 2010 was more than ten times the inflow in 1995 (Table 11.2) perhaps reflecting a more favourable and investor-friendly policy environment. However, investment in agricultural water management, whether through foreign

Table 11.2 Inflows of foreign direct investment and official development assistance (ODA) received during the period 1995 and 2010 by selected recipient countries in Sub-Saharan Africa

Country	Inflows of foreign direct investment (US$ million)[1]				Net ODA received per capita (US$)[2]
	1995	2000	2005	2010	2010
Burkina Faso	10	28	34	37[a]	68.8
Ethiopia	14	135	265	184[a]	46.1
Kenya	32	127	21	133[a]	44.7
Niger	16	9	30	947[a]	30.7
Rwanda	2	8	14	42	93.5
Sudan	18*	392	2305	2682[a]	54.1
South Africa	450*	888	6647	1553	21.8
Tanzania	150	463	494	700[a]	67.1
Uganda	121	254	380	848	54.6
Zambia	97	122	357	1041	98.1
Zimbabwe	118	23	103	105	58.8
Sub-Saharan Africa	3485	5364	25,924	38,114	47.1

*: 1991–1996 annual average; [a]: estimates; [1]: UNCTAD 2001, 2003, 2011; [2]: FAO 2012

or African funding, has been declining and is only a small proportion of that for the water sector as a whole. For example, the African Development Bank lending for agricultural water management over the period 1968–2001 was US$630 million; that is, only 14 per cent of its lending to the water sector as a whole (US$4,574 million; World Bank, 2008). Capital and investment in Sub-Saharan African agriculture (in million US$), not necessarily (all) related to direct foreign investment, increased from 287.5 in 1990 to 422.2 in 2007; that is, an annual growth of 1.9 per cent in 2000–7, coming from 2.8 per cent in 1990–99 (FAO, 2012). Just over 50 per cent of the capital stock concerns fixed assets for livestock and 25 per cent includes land development and improvements (fences, ditches, drains, etc.; FAO, 2012).

It is difficult to get a clear picture of the distribution of investments in water-related technologies for productivity improvement between the private sector and the public sector and also between rainfed agriculture and irrigation agriculture. The overall estimates are often biased towards irrigation agriculture and the public sector. For the financing of investments, the commercial farmers tend to use a combination of loans from banks and other specialized credit institutions, whereas small farmers are more likely to make use of informal credit sources and micro-finance to supplement their own inputs in cash or in kind (WG, 2006).

The bias towards the public sector and irrigation is also apparent where it concerns information on the rate of returns on investments. Whereas there were many failures in the 1970s and 1980s, recent irrigation projects have generally had acceptable rates of return (World Bank, 2008). A review of 45 donor-financed projects in Sub-Saharan Africa from 1970 onwards showed poor outcomes for projects in the 1970s and up to 1984, with investment mainly aimed at the development of new large-scale irrigation with very high costs per hectare and low or negative rates of return (Inocencio et al., 2007). After 1985, the outcomes improved: of the 22 Sub-Saharan Africa projects, all except one had an economic rate of returns ranging from 10 per cent up to 60 per cent and above (Inocencio et al., 2007; World Bank, 2008).

Policies related to water resource management

Water, like land and other natural resources is at the centre of Africa's development and socio-economic growth. Water is 'life'. When it comes to consideration of water from a policy perspective, then water is furthermore a finite and vulnerable public good. Water use, for agriculture as well as for other uses is an intrinsic part of both environmental protection and sustainable development and is key to the attainment of the Millennium Development Goals. The integrated water resource management (IWRM) strategies thrust of the 1990s was the time that across Sub-Saharan Africa countries with support from various bilateral and multilateral institutions undertook extensive reviews and design of national water policies and strategies.

In many countries, the existing national water policies or major components of them are still as originally developed within the IWRM context. The rationale for an overarching policy framework and specific policies and policy guidelines for a coherent and systematic policy environment on water (and agriculture water) is still appreciated in national development systems. However, the question is does Sub-Saharan Africa have national water policies which will enable placing water at the core of the development agenda and specially help leverage investment financing into agriculture water? In dealing with this question, two issues, which are two sides of the same coin, are: (a) quality or relevance of the policies and (b) implementation framework and capacity.

Relevance of policies

While many of the water policies in the 1990s were oriented towards ensuring integrated approaches across sector objectives and environmental conservation and resilience, more recent reviews (e.g. Savenije and van der Zaag, 2002; Veettil et al., 2011; Medellin-Azuara et al., 2012) are oriented towards aligning 'water as an economic resource' and therefore report more on appropriate and effective instruments to optimize the value of water in economic sense. Effects of climate change on water availability have also received increased attention in recent policy reviews (e.g. Bates et al., 2008; Hanjra and Quresgi, 2010). In many national water policies, components associated with industrial use including large-scale irrigation are better articulated with associated support structures for implementation.

An element linked to the quality of the policies in a country is the policy design system. One of the outstanding challenges in agriculture and water policies in Sub-Saharan Africa is the existence of institutional frameworks and supportive legislation to enable evidence-based and inclusive policy design. Mandates and responsibilities for some key components in water management policies remain fragmented in different sectors, ministries and interest groups, without clear (especially implementation-oriented) arrangements for synergies and complementarities. On the other hand, most policy design processes are not adequately supported by data, let alone quality data and the design processes do not always take the time to ensure dialogue and establishing shared vision among the implicated players and stakeholders. The lack of clear and credible data and tracking mechanisms has in many cases undermined intentions and plans on objective dialogue and accountability.

Implementation capacity and regulatory framework for policy implementation

An important consideration in policy implementation concerns the inter-institutional arrangements as well as the within-institution capacity and competencies for effective execution of policies including credible decision-making processes. This is generally the most critical issue facing national water policies. Success factors range from political will and leadership, incentives for multi-partner implementation arrangements, to shared vision including clarity in aligning towards national development objectives. Even in cases where competencies exist, the institutional and policy framework may not provide for optimal use of these areas of expertise.

Many reviews and analyses have identified lack of implementation capacity as the primary and fundamental issue that will need attention to ensure appropriate and effective national water policies. A major challenge remains building and sustaining multi-partner arrangements for implementation of the policies in a systematic and predictable way, with the roles of government and other stakeholders clearly defined. In many cases, the weakness has been in the mechanisms, the ability and quite often the political will, to track progress and performance and take necessary measures to ensure policies are upheld or, where necessary, revised.

In general, much of the work on IWRM strategies was done outside national development planning processes – arguably as a consequence of the external donor-driven nature of the process. The IWRM plans and strategies did not go further to ensure appropriate implementation capacity, desired financing as well as incentives (or dis-incentives) for implementation of the policies across the complete value chain.

National water policies and investment financing in agricultural water

National level policies, legislation and regulatory framework are crucial in attracting and expanding sustainable financing for agriculture water initiatives, whether smallholder or large-scale commercial initiatives. The policies and a predictable policy design system have direct impact on building investor confidence and management of risks, which are key aspects in attracting private sector investment financing in particular.

With the 'public good' inclination on water and especially the thrust on smallholder agriculture water use, it is crucial for governments to ensure that policies are supportive of a balanced situation with need to attract commercial private sector investment financing, on one side, and socially acceptable development initiatives, which empower smallholder in terms of access to water for agriculture purposes, on the other side.

The fragmentation in public water management investments has been undermining the power of public sector financing to leverage private sector investment. National water policies have a clear and important responsibility to economic pricing and associated accountability while at the same time ensuring that less-advantaged sections of the communities have access to water especially as an input in their economic activities. Ensuring desired flow in investment financing in agriculture water, a balanced set of policies and institutional reforms should be sought to harness the efficiency of market forces and at the same time strengthen the capacity of governments and other national stakeholders to effectively deliver on their roles.

Specifically in terms of attracting investment financing into agriculture water, two key factors will need to be considered at national policy level. These are:

(a) The largely smallholder character of farming in relation to viable business on one hand and poverty alleviation and food security objectives, on the other hand. There is a need for governments to objectively consider deliberate measures to stimulate and facilitate economic use of water in smallholder farming systems. This includes support to agriculture water use in enterprises that within a value chain consideration are able to offer viable business prospects.
(b) Agricultural water use is closely related to land access and use. Therefore, in addition to water policies, land use policies (including associated legislation) must take into account appropriate provisions for effective and efficient, as well as environmentally friendly, water use. Moreover, both national land and water management policies should ensure appropriate complementarities with existing land use policies.

Policy strategy considerations for water in agriculture

Among the strategies available to policy designers to ensure efficient water allocation and use and create a facilitating environment for financing investments are those encompassing, the fields of water legislation, water and land use rights, decentralizing water management, and pricing and market access, which will be discussed below.

Water legislation, use and access rights

The wide-ranging water acts for the selected countries in Sub-Saharan Africa in Table 11.3 often ascribe the property rights of water to the state (e.g. Tanzania), its president (e.g. Zambia, Zimbabwe), the public (e.g. Ethiopia, Rwanda) or to the private owners of the land (e.g. Mauritius), under specific circumstances. In cases where it is declared a public commodity, the grant of concessions or licenses to individuals is addressed (IELRC, 2012). A state agency is typically charged with the task of surveying the status of water resources (monitoring water resource quality and quantity) and dealing with the administration associated with licences and concessions. The newer legislations are clearly more strict in the protection of water resources, not only by emphasizing the sustainable management of the resource, but also by punishing more severely any pollution and other forms of human impact on the resource. Further, several acts explicitly acknowledge that the state or the president has the duty of granting access to freshwater to the population (see also IELRC, 2012).

Whereas some progress has been made in water legislation over the past two decades, governments in Sub-Saharan Africa are faced with considerable challenges ahead under current and future conditions of climate change, food demands, and market globalization. While some countries make use of more

Table 11.3 Water law instruments and year of enactment for selected countries in Sub-Saharan Africa

Country	Water law instrument	Year
Botswana	Water Act	1976
Burkina Faso	Law providing the orientation for water management	2001
Cameroon	Water Regime Law	1998
Cape Verde	Water Code	1984
Ethiopia	Water Resource Management Proclamation	2000
Kenya	Water Act	2002
Lesotho	Water Resources Act	1978
Madagascar	Water Code	1999
Mauritania	Water Code	2005
Mauritius	Water Resources Master Plan in preparation	-
Namibia	Water Resources Management Act	2004
Niger	Decree 93-014 Water Regime	1993
Nigeria	Water Resources Management Decree	1993
Rwanda	Water Law*	2008
South Africa	National Water Act	1998
	Water Services Act	1997
	Water Services Act – Regulations Relating to Compulsory National Standards and Measures to Conserve Water	2001
Sudan	Water Resources Act	1995
	Groundwater and Wadis Directorate Act	1998
Swaziland	Water Act	2003
Tanzania	Water Act	1974
	Water Resources Management Act	2009
	Water Supply and Sanitation Act	
Uganda	Water Statute	1995
Zambia	Water Act	1997
Zimbabwe	Water Act	2003

Source: ielrc.org/water – International Environmental Law Research Centre 2012; *: Water wiki, http://waterwiki.net

advanced policy frameworks and legislation systems others still have a long way to go, with a country like Mauritius having only a water resources master plan in preparation and lacking an official national water policy.

Instruments related to water legislation are particularly discussed within the framework of irrigated rather than rainfed agricultural systems, given the large quantities of water and the need for licenses, permits and tariffs or pricing systems that are often associated with the latter systems. Yet, also in rainfed agriculture, the capture and storage of water on land in combination with use for commercial crop production may be bound by legislation depending on the quantity of water captured, the size of the storage infrastructure, the upstream–downstream effects, and the rules and regulations of specific water law instruments.

Water rights refer to an overarching concept that embraces various aspects defining access to water, including the duration and quality of the entitlement right and its transferability (Veettil et al., 2011). Improved access to water is essential but not sufficient for sustained poverty reduction. Hanjra et al. (2009a, 2009b) call for simultaneous investments in agricultural water, education, markets and related policy support measures for reducing poverty in smallholder agriculture. Investments are also needed in policies and institutions, while lifting global agricultural trade barriers (Namara et al., 2010). Quite often a lack of secure rights to (irrigation) water is coupled to a lack of land tenure security either at individual or water user association level, suggesting the need to also address land tenure rights in order to create a facilitating environment for investment in water for agriculture (see next section).

Devolution and decentralization, democratic governance, local participation

Water in agriculture performs different functions and serves a wide range of stakeholder groups. The linkages among these functions and multiple stakeholder groups are well addressed in the IWRM approach that emerged during the 1990s (see also 'Policies related to water resource management'). This integrated approach is aimed at harmonized development and management of water, land and related resources while safeguarding ecosystem services. It is usually connected to decentralized management of water resources with local user-group participation in designated watershed areas (Ostrom, 1990). Interested and affected stakeholders serve as efficient channels through which to address socio-cultural and ecological needs where IWRM is to be put into practice.

The World Bank (2008) refers to two forms of decentralization generally taking place at government level; that is, 'decentralized sectoral' and 'decentralized local government'. The former refers to the delegation of tasks and responsibilities of sectoral ministries in the field of budgeting, coordination and implementation of activities from their national to their local level (i.e. provincial and/or district) staff. In the case of a 'decentralized local government', a proportion of the public (government and/or donor) sectoral funding is managed by local authorities such as local government units and utilized through locally prepared plans.

While decentralization could enhance the development impact of agricultural water investments, it is not a panacea for water management issues in agriculture. It is rather a means to empower rural people, providing them with facilities to develop their skills and knowledge required to participate in local political processes and hold government and private service providers accountable to them. However, the main principles on which decentralization and approaches of IWRM are based, such as the participatory involvement of users, planners and policy makers at all levels, are often not adhered to. For example, water user associations are given more delegated responsibilities whereas these are often not accompanied by sufficient delegation of power to enable them to function effectively under the new management conditions. Moreover, governments are often faced with inadequate information and lack the resources, both financial and human, to ensure effective functioning and management of water institutions. Decentralization needs strong leadership to guide the process forward, accompanied by support programmes to assist democratic governance development, build up capacity and promote empowerment.

Land distribution and land rights

Like water rights, land use and land property rights play a role in facilitating investment in water for agriculture. Yet, the type of impact that land distribution and land rights have on the investment in agriculture, whether including or excluding water-related technologies, is debated in various sources in the literature. For example, Fenske (2011) refers to the often confusing and contradictory results of studies investigating the relationship between land property rights and agricultural investment in Africa. Although no specific reference is made to investment in water-related technologies in agriculture, Fenske's study based on nine datasets from West Africa shed more light on the relationship between land tenure and agricultural investment. It revealed that land tenure affects incentives to leave land fallow and to plant trees, but it has a weaker impact on labour use and other, short-term inputs such as manure or chemical fertilizers. Jayne et al. (2010) pointing at the steady decline in the ratio of arable land to agricultural population in countries like Ethiopia, Kenya, Mozambique, Rwanda, Zambia and Zimbabwe, question the future for small farms in Sub-Saharan Africa if governments fail to provide for supportive policies and investment decisions. The latter is based on observations that land distribution patterns in the afore-mentioned countries are such that they constrain crop technology development and input intensification (with or without irrigation) because of which farmers are unable to produce more than a marginal surplus and unable to participate in commodity markets.

Tradeable water rights and water pricing in agriculture

Establishment of tradeable water rights could play an important role in improving the efficiency, equity and sustainability of water use in developing countries, with the proviso that strategies like these are more easily applicable to water in irrigated agriculture rather than in rainfed agriculture. Rosegrant and Binswanger (1994) refer to well-defined tradeable water rights as those that 'formalize and secure the existing water rights held by water users, economize on transaction costs, induce water users to consider full opportunity costs of water, and provide incentives for water users to internalize and reduce many of the negative externalities inherent in irrigation'. Yet, there is still a need for policy makers to pay more attention to institutional requirements and feasibility of developing markets in tradeable water rights. This is certainly true for regions like Sub-Saharan Africa.

Raising water prices seems to make sense in that it may encourage stakeholders to use water more efficiently and generate revenue to maintain existing water infrastructure and build new facilities where needed. However, one of the major barriers to attracting funds for investment in water for agriculture is the universal fact that water is free unless used for irrigated agriculture. In the latter case, water is often greatly under-priced with payments well below the costs of infrastructure operation and maintenance, with public irrigation institutions often heavily dependent on large subsidies (WG, 2006). Moreover, revenue collections systems are usually poorly developed and enforcement of regulations is often absent.

Because of perceived political risks and concern that higher prices would hurt poor farmers and consumers, there have been few attempts to implement and increase water prices. The irony is however that in most cases the poor suffer from subsidized water prices, with disproportional amounts of the water subsidies going to the better off; that is, urban water users and irrigating farmers (Rosegrant et al., 2002).

Water pricing is a complex matter, given the many different functions of water in agricultural systems (e.g. crop cultivation, livestock rearing, aquaculture hydropower, domestic water supply and sanitation to farmers' households). There are various water pricing methods, the preferred method being dependent on conditions such as water rights and system of local water governance. Veettil et al. (2011) investigated preferred water pricing methods among farmers in India under different water rights, price levels and system of local water governance. They found that under conditions of improved water rights farmers show increased preference for volumetric pricing, whereas this is not the case where water user associations exist.

Market access

Unless government policies and investment decisions are supportive, there is no prospect of sustainable intensification in rainfed agriculture in Sub-Saharan Africa. Jayne et al. (2010) go even further in their conclusion stating that there is no future for small farms in this region based on observations of, amongst others, farmers' sensitivity to higher grain prices and farmers' inability to participate in commodity markets, with a small group of relatively influential smallholders controlling the marketed agricultural surplus. The policy agenda for marketing and trade has changed from the early days of the green revolution (Birner and Resnick, 2010), with the liberalization of international trade and domestic markets allowing private enterprises to become increasingly important with a consequent restructuring of supply chains (Hazell et al., 2010). As a result, small farmers face increasing difficulties and costs associated with accessing markets, inputs and financial services. This calls for innovations in institutions to overcome market failures and stresses the need for enhanced collaboration among farmers, private companies, and NGOs, with ministries of agriculture and other public agencies taking on new, more facilitating roles (Hazell et al., 2010).

Linking up with the private sector

Most national water policies identify and have treated water as a public good and consider access to water as a human right. In many cases (e.g. Kenya) the national water policy does state that 'water will not be privatized'. However, many of the national water policies do identify and allude to the fact that the private sector has a role to play especially in water supply support services. The private sector has been active in issues related to portable water supply in urban centres and in industrial water supply. In these instances, water is simply another commodity and governments have normally lagged behind on regulatory functions.

Most national water policies have not deliberately and proactively pursued private sector participation especially in agricultural water use and management interventions, except in large-scale commercial farming (irrigation) enterprises involving large dams and large-scale extraction from river or underground sources. Deliberate policy incentives to encourage and attract private sector into smallholder agriculture water management systems, in terms of implementation capacity, (new) knowledge and financing is almost non-existent. As a result many interventions supporting scaling up of smallholder innovations including water lifting, pumping and efficient use have been limited and most could not be sustained beyond the project life.

Smallholder irrigation enterprises can and have proved viable businesses. Asia has demonstrated widespread and viable smallholder irrigation and agriculture water use enterprises. However, it is clear that similar developments are needed for enterprises based on rainfed agriculture making use of water harvesting systems and other related technologies. In addition, some incentives in terms of public policy are desired and sometimes essential to establish the foundation for private sector investments. Appropriate policies are essential to ensure an environment with minimum risks to the investments also considering that many people that may be the target may not have the levels of collateral required in commercial lending and that the interest rates may not make economic sense to them. These are also circumstances where the leveraging power of public sector financing would show great value.

Some policy areas where government can ensure desired incentives for private sector investments in agriculture water include:

- Predictability and reliability of the policy design system. This includes ensuring an inclusive policy design system with both state and non-state stakeholders playing an active role as well as ensuring that the process relies on credible data and evidence based analysis.

- Public sector investments in associated infrastructure including roads, dams and irrigation and water harvesting structures.
- Public sector investment in training and capacity development for farming communities on effective and modern water use technologies including improved irrigation systems. This includes training in associated land use and livestock practices which maximize water use efficiency.
- Subsidized, loan provisions, hence affordable interest rates for increased number of farming communities especially smallholder farmers.

Even with the poverty alleviation and food security objectives key in the rationale for various agricultural water use strategies, policies and interventions, it is important that wealth creation and economic growth objectives are clear in these interventions. This will ensure that, for example, irrigation and water harvesting support interventions are designed within a comprehensive and integrated system in which commodities, processing, and markets are appropriately considered.

The way forward

Given the increasing pressures on water resources and increasing demands for crop products (i.e. food, fibre and energy), the world must increase crop production whilst simultaneously reducing water allocations to agriculture (i.e. more crop per drop). The issue of water scarcity is generally less severe in Sub-Saharan Africa than in some other regions (e.g., MENA, South Asia), but climate change predictions indicate that the 600,000 km^2 currently classified as only moderately water constrained will soon become severely limited (Kijne et al., 2009: p. 34).

The potential for irrigation in Sub-Saharan Africa is less than 4 per cent of the region's total agricultural land area (*op. cit.*). Hence, expanding the current area irrigated and bridging yield gaps in irrigated agriculture cannot meet this challenge. Agriculture-driven growth will be assured and livelihoods will be improved more effectively by investing in efforts to upgrade rainfed agriculture.

There is a vast untapped potential in rainfed production systems in Sub-Saharan Africa that can be realized through knowledge-based innovations in land and water management (Kijne et al., 2009). For smallholder farmers, dry spells reduce yield and increase risk. Water harvesting sits within the array of such innovations and provides an alternative route to unlocking this potential by enabling farmers to supplement inadequate and/or unreliable direct rainfall.

Closing the yield gap and delivering sustainable intensification is not just a matter of transferring better technologies to farmers. It is essential also to put in place the institutional structures to support the adoption of improved technologies and practices. National and local policies can be supportive or provide barriers and disincentives to the adoption of innovations. Targeted policy actions are required to support sustainable intensification of rainfed agriculture. Support in terms of creating a policy environment for encouraging investment in water harvesting for smallholder agriculture in the drylands of Sub-Saharan Africa is a must.

A cautionary note must be sounded. Investment in water harvesting will deliver benefits at local level to individual farmers and rural communities. However, extensive development of water-harvesting structures may have negative consequences for downstream water resources. In effect, blue water is being converted to green water by these structures. Evidence exists in India where large-scale adoption of water harvesting structures has had this effect (Kijne et al., 2009). Any plan based on improved rainfed agriculture and increased use of green water brings trade-offs in that there may be less blue water for downstream users and for environmental functions.

In a water scarce environment it should be recognized that any land-use decision is also a water-use decision. Increasingly, scientists are realizing that the consideration of green water flows in water resources management will require a higher level of integration which explicitly considers land issues together with water issues, sometimes termed Integrated Land and Water Resources Management (ILWRM). It has been suggested (Jewitt, 2006) that ILWRM may require policies that are more explicitly directed at the management of the land resource as already exist in the context of management of 'streamflow reduction activities' in South Africa.

References

Bates, B., Kundzewicz, Z. W., Wu, S. and Palutikof, J. (2008) 'Climate change and water', IPCC Technical Paper VI, available at: http://www.ipcc.ch/pdf/technical-papers/climate-change-water-en.pdf, accessed 20 May 2012.

Birner, R. and Resnick, D. (2010) 'The political economy of policies for smallholder agriculture', World Development, vol. 38, no. 10, pp. 1442–1452.

FAO (2003) Unlocking the Water Potential of Agriculture, Food and Agriculture Organization of the United Nations (FAO), Rome.

FAO (2005) Summary of World Food and Agricultural Statistics 2005, Food and Agriculture Organization of the United Nations (FAO), Rome.

FAO (2010a) FAO Statistical Yearbook 2010, Food and Agriculture Organization of the United Nations (FAO), Rome.

FAO (2010b) The State of Food Insecurity in the World: Addressing food insecurity in protracted crises 2010, Food and Agriculture Organization of the United Nations (FAO), Rome.

FAO (2012) FAO Statistical Yearbook 2012, Food and Agriculture Organization of the United Nations (FAO), Rome.

FAO-AQUASTAT available at http://www.fao.org/nr/water/aquastat/main/index.stm, accessed 2 June 2012.

Fenske, J. (2011) 'Land tenure and investment incentives: Evidence from West Africa', Journal of Development Economics, vol. 95, pp. 137–156.

de Fraiture, C., Molden, D. And Wichelns, D. (2010) 'Investing in water for food, ecosystems, and livelihoods: An overview of the comprehensive assessment of water management in agriculture', Agricultural Water Management, vol. 97, no. 4, pp. 495–501.

de Fraiture, C. and Wichelns, D. (2010) 'Satisfying future water demands for agriculture', *Agricultural Water Management*, vol. 97, no. 4, pp. 502–511.

Hanjra, M. A., Ferede, T. and Gutta, D. G. (2009a) 'Pathways to breaking the poverty trap in Ethiopia: investments in agricultural water, education and markets', *Agricultural Water Management*, vol. 96, no. 11, pp. 1596–1604.

Hanjra, M. A., Ferede, T. and Gutta, D. G. (2009b) 'Reducing poverty in Sub-Saharan Africa through investments in water and other priorities', *Agricultural Water Management*, vol. 96, no. 7, pp. 1062–1070.

Hanjra, M. A. and Qureshi, M. E. (2010) 'Global water crisis and future food security in an era of climate change', *Food Policy*, vol. 35, no. 5, pp. 365–377.

Hazell, P., Poulton, C. and Wiggins, S. (2010) 'The future of small farms: trajectories and policy priorities', *World Development*, vol. 38, no. 10, pp. 1349–1361.

Hoekstra, A. Y. and Mekonnen, M. M. (2012) 'The water footprint of humanity', *PNAS*, vol. 109, no. 9, pp. 3232–3237.

IELRC (2012) 'Selected water law instruments around the world', International Environmental Law Research Centre (IELRC), available at: ielrc.org/water, accessed on 20 May 2012.

IFAD (2002) 'Desk review of the PRSP process in Eastern and Southern Africa', International Fund for Agricultural Development, Programme Management Department, Africa Division II, Rome.

Inocencio, A., Kikuchi, M., Tonosaki, M., Maruyama, A., Merrey, D., Sally, H. and de Jong, I. (2007) *Costs and performance of irrigation projects: A comparison of Sub-Saharan Africa and other developing regions*, IWMI Research Report 109, International Water Management Institute Colombo, Sri Lanka, 81 pp.

Jayne, T. S., Mather, D. and Mghenyi, E. (2010) 'Principal challenges confronting smallholder agriculture in Sub-Saharan Africa', *World Development*, vol. 38, no. 10, pp. 1384–1398.

Jewitt, G. (2006) 'Integrating blue and green water flows for water resources management and planning', *Physics and Chemistry of the Earth*, vol. 31, no. 15–16, pp. 753–762.

Kijne J., Barron J., Hoff H., Rockström J., Karlberg L., Gowing J., Wani S. and Wichelns D. (2009) 'Opportunities to increase water productivity in agriculture with special reference to Africa and South Asia', Stockholm Environment Institute, Stockholm, 39 pp.

Medellin-Azuara, J., Howitt, R. E., and Harou, J. J. (2012) Predicting farmer responses to water pricing: rationing and subsidies assuming profit maximizing investment in irrigation technology, *Agricultural Water Management*, vol. 108, pp. 73–82.

Mekonnen, M. M. and Hoekstra, A. Y. (2011) 'National water footprint accounts: the green, blue and grey water footprint of production and consumption', *Value of Water Research Report Series No. 50*, vol. 1 and 2, UNESCO-IHE, Delft.

Molden, D. (2007) *Water for Food, Water for Life: A comprehensive assessment of water management in agriculture*, Earthscan and International Water Management Institute, London and Colombo.

Molden D., Oweis T., Steduto P., Bindraban P., Hanjra M. A. and Kijne J. (2010) 'Improving agricultural water productivity: between optimism and caution' *Agricultural Water Management*, vol. 97, no. 4, pp. 528–535.

Namara, R. E., Hanjra, M. A., Castillo, G. E., Ravnborg, H. M., Smith, L. and Van Koppen, B. (2010) 'Agricultural water management and poverty linkages', *Agricultural Water Management*, vol. 97, no. 4, pp. 520–527.

OECD–FAO (2011) *OECD-FAO Agricultural Outlook 2011*, OECD Publishing, DOI: 10.1787/agr_outlook-2011-en, accessed 20 May 2012.

Ostrom, E. (1990) *Governing of the Commons: Incentives, Rules of the Game and Development*, Annual World Bank Conference on Development Economics, World Bank, Washington D.C.

PRB (2011) *World Population Data Sheet*, Population Reference Bureau (PRB), Washington D.C.

Rosegrant, M. W. and Binswanger, H. P. (1994) 'Markets in tradable water rights: Potential for efficiency gains in developing country water resource allocation', *World Development*, vol. 22, no. 11, pp. 1613–1625.

Rosegrant, M. W., Cai, X. and Cline, S. A. (2002) *Global Water Outlook 2025: Averting an Impending Crisis*, International Food Policy Research Institute, Washington D.C. and International Water Management Institute, Colombo, Sri Lanka.

Savenije, H. and van der Zaag, P. (2002) 'Water as an Economic Good and Demand management Paradigms with Pitfalls', *Water International*, vol. 27, no. 1, pp. 98–104.

SUA (2007) *Country Report: Tanzania – Managing Risk and Reducing Vulnerability of Agricultural Systems under Variable and Changing Climate*, Soil-Water Management Research Programme, Sokoine University of Agriculture, September 2007, available at: www.suanet.ac.tz/ccaa/downloads/crTanzania.doc, accessed on 10 May 2012.

UNCTAD (2001) *World investment report 2001 – Promoting linkages*, United Nations Conference on Trade and Development (UNCTAD), UN, New York and Geneva.

UNCTAD (2003) *World investment report 2003 – FDI Policies for development: National and international perspectives*, United Nations Conference on Trade and Development (UNCTAD), UN, New York and Geneva.

UNCTAD (2011) *World investment report 2011 – Non-equity modes of international production and development*, United Nations Conference on Trade and Development (UNCTAD), UN, New York and Geneva.

Veettil, P. C., Speelman, S., Frija, A., Buysse, J. and van Huylenbroeck, G. (2011) 'Complementarity between water pricing, water rights and local water governance: a Bayesian analysis of choice behaviour of farmers in the Krishna river basin, India', *Ecological Economics*, vol. 70, no. 10, pp. 1756–1766.

WG (2006) 'Financing water for agriculture', Progress report no. 1, Working Group on Financing Water for Agriculture, Contribution to the task Force on Financing Water, 4th WWF, Mexico, 2006.

World Bank (2006) *Reengaging in Agricultural Water Management: Challenges and Options*, World Bank, Washington D.C.

World Bank (2008) *Investment in Agricultural Water for Poverty Reduction and Economic Growth in Sub-Saharan Africa: synthesis report*, A collaborative program of AFDB, FAO, IFAD, IWMI, and the World Bank, World Bank, Washington D.C.

World Bank database (2012) World databank, World Development Indicators (WDI) & Global Development Finance (GDF), available at: http://databank.worldbank.org/ddp/home.do, accessed 20 May 2012.

World Resources Institute (WRI) in collaboration with United Nations Development Programme, United Nations Environment Programme, and World Bank (2005) *World Resources 2005: The Wealth of the Poor – Managing Ecosystems to Fight Poverty*, WRI, Washington D.C.

Chapter 12

Conclusions, lessons and an agenda for action

John Gowing and William Critchley

Evidence of change

In this book we have examined change over a quarter of a century from a baseline that has been defined in different chapters with varying levels of precision. We have set out to compare the situation now, with conditions that were described in various earlier sources, most notably in the Sub-Saharan Africa Water Harvesting Study (Critchley et al., 1992), and in a study of indigenous soil and water conservation in Africa (Reij et al., 1996). Other important baseline reference points include Pacey and Cullis (1986) and Rochette (1989).

Looking back 25 years, it was a time when the advocacy of participatory approaches in agricultural development was still to become mainstreamed, and the value of indigenous knowledge was barely recognized. Chambers (2008) provides an account of the evolution and spread of participatory methods over that period. Briggs (2005) reviews the increasing role of indigenous knowledge (IK) in development, and Reij and Waters-Bayer (2001) highlight the role of farmer innovation. It is clear from these and many other sources that in the past two or three decades both participation and IK have become standard components of the sustainable development paradigm, which itself has become ubiquitous. So are these changes in development theory evident on the ground in Africa, and in particular in relation to water harvesting for agriculture?

Going back to those earlier times, Pacey and Cullis (1986) pointed out that '[i]nformation about existing traditions of runoff farming is inadequate nearly everywhere' and Reij et al. (1988) remarked that there was 'little information available on water harvesting in Sub-Saharan Africa'. As shown in the overview of historical trends in Chapter 2 and the review of literature in Chapter 3, but also from evidence in the country case study reports (Chapters 4 to 10), this situation has changed significantly. The efforts of the World Overview of Conservation Approaches and Technologies in uncovering documenting and spreading knowledge (https://www.wocat.net/) deserves a special mention. The documentary evidence from various sources is also now supplemented by video imagery, available on The Water Channel (http://www.thewaterchannel.tv/en/videos/showcase). Information on traditional and introduced practices for water harvesting has expanded greatly, though the 'digital divide' still poses a serious constraint to those in Sub-Saharan Africa who have poor and slow internet connections. But has this knowledge, and awareness of barriers, led to improvements in practices of agencies involved in promoting water harvesting?

To answer that question more fully, we present a series of issues raised in the earlier days, and seek to provide enlightenment on how these have been addressed. For example it was reported by Reij et al. (1988) that 'projects do not systematically monitor the impact of their activities' and that 'there seems to be a gap between research and implementation'. Critchley et al. (1992) highlighted issues and made recommendations that they believed would, or could, lead to improvements in the implementation of water harvesting programmes. They also underscored the importance of better monitoring and evaluation. However they went further and specified the need for data collection that could lead to better cost-benefit analysis and socio-economic impact assessment. From our current perspective this scarcely seems to have improved over the years, judging from the conclusions of the literature review and a broad consensus in the country chapters.

There were other analytical points made in Critchley et al. (1992); for example, there was a flag waved about dependence on heavy machinery in construction, and the need to develop more appropriate and affordable tools. While the case from Niger described the only significant use of specialized tractors for construction (Chapter 7; see also Photos 7.1 and 7.2) this warning seems to have been fully justified in that case. However, in Sudan (Chapter 9) the use of machinery for construction of floodwater harvesting diversion bunds is inevitable, due to the size of structure required. More appropriate tools? The front-mounted blade on a small tractor in Sudan – for *teras* construction may be seen as a positive development in this sense, being cheap and easy to maintain, in a country where much agriculture is mechanized.

The same authors were concerned about fertility management associated with water harvesting: and there is plenty of evidence in this volume that the problem is being addressed, or at least is taken seriously. Apart from mention in several country chapters, the literature review in Chapter 3 cites examples to support the benefits of keeping soil fertility in mind when increasing the amount of water available to plants. 'Improved husbandry' is a recommendation of Critchley et al. (1992) that is scarcely mentioned other than in the context of intensive horticultural production when water is stored in ponds and this allows the switch to high value crops (e.g. Ethiopia, Chapter 5, from household ponds; Kenya, Chapter 6, from road runoff harvesting).

Coordination of national incentive policies was another recommendation in Critchley et al. (1992). The implicit notion here was that policies should be developed to make sure that incentives were only used to 'kick-start' activities, or permitted for a short time in situations of emergency under food-for-work programmes. Several countries hosted projects and programmes that were contradictory regarding philosophy and technologies, but in their policies over 'handouts' also. It is surprising that, while there are examples of countries indeed coordinating various initiatives (e.g. Sudan, Chapter 9), food-for-work (while now renamed 'food-for-assets') is still on the scene and appears to be accepted in several countries as a form of semi-continuous social welfare support (Ethiopia, Chapter 5, is perhaps the best example).

Land tenure has continued to surface as a problem that is widely perceived to restrain efforts at land improvement. While the *a priori* case for a link between land tenure and agricultural investment is strong, 25 years of accumulated experience still does not provide clear evidence of the role it plays in determining the success of water harvesting interventions. The existing literature on the relationship between property rights in land and agricultural investment in Africa is often confusing and contradictory (Fenske, 2011). Land tenure does not generally affect short-lived investments and inputs such as labour, fertilizer, manure and pesticides. For longer term investments such as tree planting and water harvesting there are countries where rights are seen to be conditional on use. New analysis and greater clarity is still required.

A number of other issues raised or recommendations made have been addressed to a greater or lesser extent: the involvement of women within decision making has become an accepted part of the 'participatory approach' (as mentioned earlier). National institutional responsibility and coordination of efforts certainly seems to have emerged to a greater or lesser extent: the days of myriad NGOs free to do as they choose is drawing to a close.

Bright spots are evident where water harvesting techniques have been successfully introduced and/or adapted. Water harvesting seems to have made the most progress where systems based on traditions have been stimulated and supported (e.g. in Burkina Faso, Chapter 4; Niger, Chapter 7; and Tanzania, Chapter 8); where national programmes have begun to take water harvesting seriously (most of the countries revisited, barring Zimbabwe); where international cross visits have been organised (e.g. Rwanda to Kenya, Chapter 6; and Niger to Burkina Faso, Chapter 7); where local innovation has led to the evolution of new technologies (e.g. Kenya, Chapter 5; and Zimbabwe, Chapter 10) and finally where commercial models are being developed (e.g. Ethiopia, Chapter 5; and Kenya, Chapter 6).

Thus while there have been developments they have been sporadic in terms of specific elements of water harvesting, and differ from country to country. Many of the recommendations made decades ago are still as valid today as they were before: we do not need to completely reinvent the wheel. It is time to revisit these recommendations, learn from progress where it has been made – and above all to formulate plans for action. If water harvesting was important then, it is more so now as we enter unknown territory of increasing population pressure and uncertainties regarding climate. Thus we propose an 'agenda for action' based on experience and the lessons drawn.

An agenda for action

The following, developed from Scheierling et al. (2012), sets out our views on the priorities to take water harvesting forward this century. It is firmly embedded in the conclusions and lessons we have presented in the foregoing section.

Proceed with confidence

Despite the various barriers to wider adoption, there surely is a way forward. In fact there are more examples of success, than are often acknowledged. This analysis has underscored the lack of performance and impact data – and has strongly recommended improvements in monitoring and evaluation, as well as research, into impact and spread of various water harvesting technologies. We have now enough experience of technologies, and the approaches that create an enabling environment, to go ahead with a vigorous but strategic campaign of promotion – accepting that land users will have the final say.

There is increased evidence that water harvesting has considerable potential to improve productivity based on better runoff and fertility management; and it has the ability to provide more secure food production and improve livelihoods. In some situations it can lift the rural poor out of their poverty. It is time to recognize that water harvesting in Sub-Saharan Africa has come of age: it must no longer be seen solely as the 'work' element of 'food-for-work' but as a valid form of sustainable land management in its own right. There is a clear role for water harvesting, just as there is in the more humid areas for *in situ* moisture conservation systems such as conservation agriculture. Time, it might be said, for water harvesting to be championed just as agroforestry, the system of rice intensification and conservation agriculture all are.

Promote proven technologies

Chapter 2 (Table 2.3) presents a compilation and summary of twelve technologies that either have spread spontaneously and performed strongly, or at least have potential merit, and are acceptable in given situations to certain farmers. Some are local innovations, others have been introduced from outside, but many have their foundations in local traditions. Naturally none of these is ubiquitously appropriate: just as water harvesting is no panacea, none of these technologies should be given undeserved accolades. Some of the technologies lend themselves to hand-based systems, others need mechanization; some depend on a local supply of loose stone; they differ in terms of agro-climatic and edaphic suitability; the crop or pasture or trees to be planted will also help define suitability of system. There may be no obvious logic why one system is preferred by local farmers to another. So what is the main message regarding technologies? This is: promote water harvesting technologies that we know have worked in similar conditions, but allow land users the ultimate choice.

Support knowledge management systems

It is a tenet of WOCAT, and is surely self-evident, that much knowledge exists locally about water harvesting in Sub-Saharan Africa. Yet, still, the knowledge hasn't been fully uncovered and hasn't reached the places and the people where it can make best impact. Knowledge management, from uncovering tacit information to disseminating it through mechanisms that reach receptive target groups is of paramount importance. Language barriers – particularly the francophone/ anglophone divide in Sub-Saharan Africa – must be broken down. There are many ways of achieving these objectives. It includes support to global information networks (like WOCAT) that emphasize the search to document hitherto hidden 'know-how'. Surely there are various methods of making use of the mobile phone to share ideas and spread knowledge in Africa. Let us determine which are the most effective. There again there is a school of thought that says much information is already known and documented: but is itself hidden in 'grey' literature, which is buried even deeper as students and researchers rely more and more on the internet. And there remains the age-old principle that travel broadens the mind. Simply moving farmers around and letting them see for themselves is a powerful way of spreading knowledge.

Build up our knowledge base

It remains true that there are forms of water harvesting that are unknown to the scientific world. They may be traditions, or (more likely) evolving innovations. The search for new systems, or variations on technologies, must not stop. There is certainly inadequate information on impact of water harvesting: this has been referred to frequently in this book. It is high time that monitoring and evaluation is improved, and research into specific impact areas initiated. There are questions about production, about cost-benefits, as well as impacts on livelihoods. Furthermore we have scanty knowledge about environmental impacts: with increased use of water harvesting – and thus re-routing significant flows – there will be important upstream–downstream interactions that may need to be addressed. However caveats need to be introduced here. First, monitoring and evaluation is a burden, thus it should be used strategically and sparingly: how much do we really *need* to know? The concept of 'optimal ignorance' needs to be applied in this case. How can the process be made useful and appealing to the monitor him/ herself? Second, and related to the forgoing, we should beware of research for the sake of research. Third, something that has been clear during the exercise of revisiting the old sites and has been alluded to above; much has been researched and written about already, and some 'old' writings are well worth revisiting.

Stimulate innovation systems – especially local innovation

There is increasing discussion about innovation systems and how to stimulate these to function better. An innovation system comprises actors from different constituencies (research, extension, private sector and academia) involved in various activities (research, product development, promotion) that provide new ideas and products. There is, however, another dimension to innovation systems, and that is the stimulation of local adaptation and innovation. If the subject is water harvesting, then farmers have driven the development of such systems for millennia: it is time to recognize this and harness their role in local, adaptive innovation. Indeed, they are continuing to do so, as we have seen in the various country chapters: not only in terms of technology, but also developing 'business models'. Ironically, climate change may very well be a positive stimulant to local innovation in water harvesting. Naturally, researchers need to be brought into the process to assist farmers to develop and monitor systems.

Broaden our outlook through acknowledging 'Water Harvesting Plus' (WH+)

Water harvesting has too often been sold purely on its technical objective: the aim of increasing the amount of rainfall runoff for production. But as we have seen, it is much more than that. There are soil fertility issues: by removing the first limiting factor of water, we immediately uncover the second, and commonly that is soil fertility. Crop choice is crucial. Aspects of suitable plants may be drought tolerance (in case of inadequate rainfall and runoff), but also waterlogging tolerance (to deal with temporary inundation) and/or drought evasion (to grow quickly on residual moisture). Farmer knowledge is important here, as is research assistance. Other aspects of agronomic practices that must be stressed in water harvesting are weed control, mulching, tillage, and intercropping – including agroforestry mixes. Water harvesting also needs to be seen in the light of reducing land degradation and improving ecosystem function. There are socio-economic factors involved: livelihoods, and in some situations social welfare is closely connected to water harvesting. Thus to move forward, water harvesting needs to considered, and sold, in the broadest light of what we term 'water harvesting plus' (WH+).

Promote water harvesting as an ally against climate change

Integral within the concept of WH+ is the ability of water harvesting to function as a constituent of climate-smart agriculture in the semi-arid and arid regions. It can contribute to the 'triple-win' by building up system resilience, by helping to store carbon, and most obviously by improving production. Water harvesting can also be viewed as a climate adaptation technology for two reasons: through stabilizing production under climate-change induced drought, but also by being adaptable itself by land users, who can adjust the catchment area and also react to rainfall variations through 'response farming': thus adapting the system structurally and agronomically.

Encourage an enabling environment

An enabling policy framework – with relevant policies and means of implementing them – is crucial. In WOCAT's analysis of successful soil and water conservation projects (WOCAT, 2007), the need to break out of the conventional three-year project mode and move towards programmatic approaches that involve various partners is stressed. Another key is to establish appropriate incentive mechanisms. Caution needs to be employed to avoid the hand-out dependency syndrome where stimulants become addictive. One theme that was stressed earlier (Chapter 3) was the role of the market and the hindrance caused by 'missing markets'. Unless land users can get access to inputs and are able to market their produce, then this will deter them from moving beyond subsistence production. This is a key area where policy makers can make a difference.

Fulfill our obligations

Finally – for reasons simultaneously involving poverty and the environment – there is a national and international obligation to invest in rainfed agriculture – whether water harvesting or systems of conservation agriculture – or other forms of sustainable intensification. The link between climate change and sustainable land management through the huge potential for carbon sequestration in the land, and the enormous amounts of carbon lost through land degradation, should help to focus international attention on what has too often been seen as a local problem. Technology is not the main limiting factor: it is the willingness of governments and development agencies to invest in terms of money for research and development, alongside concerted implementation efforts over a prolonged period – based on lessons from experience – that is principally lacking.

References

Briggs, J. (2005) 'The use of indigenous knowledge in development: problems and challenges', *Progress in Development Studies*, vol. 5, no. 2, pp. 99–114.

Chambers, R. (2008) *Revolutions in development enquiry*, Earthscan, London.

Critchley, W. R. S., Reij, C. and Seznec, A. (1992) 'Water harvesting for plant production, Volume II: Case studies and conclusions for Sub-Saharan Africa', World Bank Technical Paper Number 157, Africa Technical Department Series, Washington D.C.

Fenske, J. (2011) 'Land tenure and investment incentives: Evidence from West Africa', *Journal of Development Economics*, vol. 95, pp. 137–156.

Pacey, A. and Cullis, A. (1986) *Rainwater Harvesting: the Collection of Rainfall and Runoff in Rural Areas*, IT Publications, London.

Reij, C., Mulder, P. And Begemann, L. (1988) 'Water harvesting for plant production', World Bank Technical Paper 91, World Bank, Washington D.C.

Reij, C., Scoones, I. and Toulmin, C. (eds) (1996) *Sustaining the Soil*, Earthscan Publications, International Institute for Environment and Development, London.

Reij, C. and Waters-Bayer, A. (eds) (2001) *Farmer Innovation in Africa*, Earthscan, London.

Rochette, R. M. (1989) *Le Sahel en Lutte Contre la Desertification: Leçons d' experiences*, GTZ, Eschborn.

Scheierling, S. M., Critchley, W. R. S., Wunder, S. and Hansen, J. W. (2012) 'Improving water management in rainfed agriculture: Issues and options in water- constrained production systems', Water Paper, Water Anchor. The World Bank, Washington D.C.

WOCAT (2007) *Where the land is greener: case studies and analysis of soil and water conservation initiatives worldwide*. Liniger, H. and Critchley, W. (eds), CTA, FAO, UNEP, CDE, Bern.

Glossary

Adoption   Voluntary uptake of a system by farmers/villagers

Agroforestry   Combination of trees or shrubs with crops or livestock to provide synergies

Agropastoralism   Combination of pastoralism with cropping

Bund   Earth or stone structure larger than a ridge

Caag   Traditional WH system in Somalia where earth bunds impound channel flow

Catch crop   Crop (usually rapidly-maturing) planted to make use of (unexpectedly) favourable conditions

Catchment   Area of land surface producing runoff (also 'catchment area')

Catchment: Cultivated area ratio (C:CA ratio)   Runoff generating catchment area relative to cultivated area

Conservation agriculture   An *in situ* conservation/production system with the three principles of (i) minimum soil disturbance (ii) permanent organic soil cover and (iii) diversification/rotation of crops

Contour   Line joining to points of the same elevation on the land's surface

Contour bund   Earth or stone structure, larger than a ridge, following the contour

Contour furrow   Broad low contour ridges with wide furrows in-between and earth ties to prevent lateral movement of water. Ridge is catchment and furrow is cultivated area (from Zimbabwe)

Contour ridge   Small earth structure, usually 10–20 cm in height, which follows the contour

Cut-off drain   Trench made to protect cultivated land from external runoff

Demi-lune   French term, literally 'half moon', for semi-circular bund

Dead-level contour   Level contour bund with trench upslope designed to capture runoff from within/outside field: constructed in series. Infiltration ditches may be excavated within the trenches (from Zimbabwe)

Digue filtrante   French term for permeable rock dam

Ephemeral flow   Flow which occurs for short duration, often in torrents (or 'spate'), in normally dry watercourse

External catchment system   WH technique making use of overland flow (sometimes rill or channel flow), from catchments outside the cropped area

***Fanya juu* terrace**   Earth terrace bund/bank formed by throwing soil upslope (from Kiswahili: 'do up')

Floodwater harvesting (or water spreading)   WH technique making use of floodwater flow from an ephemeral watercourse

Food-for-assets   Food rations supplied in exchange for labour aimed at creating 'assets' useful to the beneficiaries (new term for food-for-work)

Food-for-work   Food rations supplied in exchange for labour (original term)

FYM   Farmyard manure

Green water   Water held in the soil and available for plant production

Horizontal interval   Ground distance between the two structures *in situ* moisture conservation systems conserving rainwater where it falls, permitting no runoff (thus differing in principle from water harvesting)

Intercropping   Combination of crops grown simultaneously in a mixture, often in alternating lines

Khor   Type of *wadi* – but usually smaller

Microcatchment   WH technique making use of overland flow from small catchments, typically a few metres in length

***Negarim* microcatchment**   Specific microcatchment technique of catchment basins used for trees, normally in a continuous block with diamond-shaped pattern

Overland flow   Runoff occurring when rainfall intensity exceeds the infiltration capacity of the soil; and before runoff concentrates in rills or gullies

Permeable rock dam   Long, low, level rock structure across a valley bottom used for water spreading

Ratoon crop   Regrowth of, for example, a sorghum crop providing a second harvest (of grain, or at least fodder) after first harvest has been taken

Relay crop   Supplementary crop planted amongst established stand

Response farming   Opportunistic management decisions taken in response to short-term climatic, or other, factors; for example, planting a relay crop

Replicability   'Transferability' of a technique from one situation to another

Ridge   Earth structure smaller than a bund. Usually 10–20 cm high spaced closer together than bunds. May follow the contour, be slightly graded, or straight

Semi-circular bunds   Ridges of earth formed in a semi-circular shape (*demi-lunes* in French)

Spate irrigation   Form of irrigation that uses floodwaters (or 'spate flow') from ephemeral or seasonal watercourses with, generally, well-developed infrastructure

Spillway   Outlet allowing safe discharge of excess runoff

Stover   Leaves and stems of crops left as harvest by-product

Tassa   Hausa word for a *zaï* – type structure used in Niger

Teras Traditional water harvesting technique from Sudan: plots bunded on three sides and open end accepting runoff from catchment

Ties (cross-ties) Earthen plugs made in furrows to prevent lateral flow of runoff

Trapezoidal bund Type of external catchment technique where runoff is impounded behind an earth bund with configuration of three sides of trapezoid when seen from above

Vertical interval Spacing between two structures determined on the basis of a fixed difference in ground elevation

Wadi Arabic term for a watercourse with ephemeral flow

Water harvesting The collection and concentration of rainfall runoff, or floodwaters, for plant production (in the context of this book)

Water harvesting plus (WH+) Water harvesting seen in an holistic framework: including agronomic, edaphic, social, economic and political aspects, etc.

Water spreading See 'floodwater harvesting'

Wingwalls Side bunds extending upslope from a base, or bottom, bund – for example in a trapezoidal bund system

Zaï Hand-dug, deep and wide, planting pits used in Burkina Faso and other parts of West Africa for concentration of runoff

Index